T0180195

Environmental Footprints and Eco-design of Products and Processes

Series Editor

Subramanian Senthilkannan Muthu, Head of Sustainability - SgT Group and API, Hong Kong, Kowloon, Hong Kong

Indexed by Scopus

This series aims to broadly cover all the aspects related to environmental assessment of products, development of environmental and ecological indicators and eco-design of various products and processes. Below are the areas fall under the aims and scope of this series, but not limited to: Environmental Life Cycle Assessment; Social Life Cycle Assessment; Organizational and Product Carbon Footprints; Ecological, Energy and Water Footprints; Life cycle costing; Environmental and sustainable indicators; Environmental impact assessment methods and tools; Eco-design (sustainable design) aspects and tools; Biodegradation studies; Recycling; Solid waste management; Environmental and social audits; Green Purchasing and tools; Product environmental footprints; Environmental management standards and regulations; Eco-labels; Green Claims and green washing; Assessment of sustainability aspects.

More information about this series at http://www.springer.com/series/13340

Miguel Ángel Gardetti · Ivan Coste-Manière
Editors

Sustainable Luxury and Craftsmanship

 Springer

Editors
Miguel Ángel Gardetti
Center for Studies on Sustainable Luxury
Buenos Aires, Argentina

Ivan Coste-Manière
Luxury and Fashion Management
SKEMA Business School
Nice, France

ISSN 2345-7651 ISSN 2345-766X (electronic)
Environmental Footprints and Eco-design of Products and Processes
ISBN 978-981-15-3771-4 ISBN 978-981-15-3769-1 (eBook)
https://doi.org/10.1007/978-981-15-3769-1

This Springer imprint is published by the registered company Springer Nature Singapore Pte Ltd.
The registered company address is: 152 Beach Road, #21-01/04 Gateway East, Singapore 189721, Singapore

Preface

Several authors have asserted that one of the luxury basics is high quality and savoir faire,[1] being the latter related to the craftsmanship of handmade products resulting from a knowledgeable elite of artisans who have preserved unique manufacturing traditions, enabling companies to continuously deliver top-quality products. The importance of craftsmanship in the luxury world should not be ignored. It was the main reason because some organizations developed important activities related to the craftsmanship in the luxury sector. The time factor is inherent to quality. Moreover, the relationship between luxury and quality makes most luxury products to be seen by luxury consumers as high price, not readily affordable and, sometimes, snobbish products. But quality is also related to unique raw materials or production processes in specific geographic areas. There is no such a thing as luxury without quality—which is its essence. Quality and craftsmanship of these iconic items help them remain desirable generation by generation, providing them a sense of timelessness. Besides, sustainable luxury would not only be the vehicle for more respect for the environment and social development, but it will also be synonym of culture, art, and innovation of different nationalities, maintaining the legacy of local craftsmanship. This volume[2] gives a comprehensive outlook about this subject and begins with the work titled "Jewellery Between Product and Experience: Luxury in the Twenty-First Century" by Prof. Alba Cappellieri, Prof. Livia Tenuta, and Dr. Susanna Testa. In this chapter, they analyse through case studies the different ways for the jewellery sector to promote a sustainable practice. It presents the opportunities and risks of progress on technological and digital innovation for the competitiveness of companies in both production processes and also explores communication tools.

---

[1]This is related to **aesthetics and beauty**: The combination of aesthetics, beauty, craftsmanship, and quality is very hard to find outside the luxury market.

[2]This volume is an effort of the **Global Center on Sustainable Luxury**, created by the agreement between **SKEMA Business School** (France) and **the Center for Studies on Sustainable Luxury** (Argentina) on 24 July 2018.

The following chapter, "Sustainable Luxury, Craftsmanship and Vicuna Poncho", written by Roxana Amarilla, Miguel Ángel Gardetti, and Marisa Gabriel, explores the theoretical framework to subsequently delve into the Vicuña Poncho universe. It offers a description of both this object, highlighting its historic and current symbology, and the ancient aboriginal practices still used by artisans, to conclude with thoughts about the importance of sustainable luxury and the appreciation of the Vicuña Poncho to empower artisans and expand their horizons.

Then, Dr. Cindy Lilen Moreno Biec develops the chapter titled "Unwritten: The Implicit Luxury". It describes and analyses how ancestral communities developed unwritten languages through textiles and how understanding, engaging with, and preserving the dynamic they used can help us connect with others in the deepest way. It explores its potentials to be considered a new form of luxury, one that is no longer about buying things but about understanding and reacting to the needs of both people and the environment in a holistic way.

Subsequently, the purpose of Matteo De Angelis, Cesare Amatulli, and Margherita Zaretti's work "The Artification of Luxury: How Art Can Affect Perceived Durability and Purchase Intention of Luxury Products" is to present and explore the idea of associating luxury products and brands with the concept of art and artworks might help luxury companies tackle "democratization" of luxury, their need to reinforce aesthetic, moral and symbolic value, and the growing consumers' concern about the social and environmental impact that luxury brands' activities bring forth.

Moving on to the next chapter, "Lasting Luxury: Arts and Crafts with Xia Bu 夏布, a Traditional Handloomed Ramie Fabric", the author, Ying Luo, present ramie and traditional craftsmanship techniques, for example, on a woman's clothing piece from Han Dynasty (206 BC–220 AD) made from natural fibre and no colour. This chapter analyses ramie fibre, presents its characteristics, and explores its potential as luxury fibre emphasizing on artisan techniques.

The following chapter entitled, "Luxury Craftsmanship as an Alternative to Building Social Fabric and Preserving Ancestral Knowledge: *A Look at Colombia*", by Alejandra Ospina and Ana López, explores the strong link between handicrafts and luxury, and between creators and creations, and the value of generating products with personal history and durability over time. Sustainability, luxury, and ethics initiate a permanent and unalterable dialogue in the reconfiguration of the fashion system. This dialogue allows addressing in a meaningful manner the creation of pieces aligned with responsible processes allowing environmental and social balance in a society where luxury cannot exist without sustainability.

Finally, Annette Condello in the chapter entitled "Crafting Luxury with 'More-ish' Qualities at the YSL Museum: An Organic Approach" analyses the "more-ish" qualities, defined herein as causing one's impulse to create *handsome* foci (as opposed to the Moorish, which is culturally based), inherent in the YSL Museum from an art–architectural perspective within the sustainable luxury context. It discusses how traditional bricklaying techniques and their unexpected connections have transformed the desert-built environment and speculates why these changes inform adaptive reuse practice as a "more-ish" organic approach. In this respect, in discussing the brick-making as a form of crafting luxury the process has cultivated a bonding or "tuning-in" tactic, important in understanding sustainable Moroccan culture.

Buenos Aires, Argentina   Miguel Ángel Gardetti
Nice, France   Ivan Coste-Manière

Contents

About the Editors

Dr. Miguel Ángel Gardetti was Renowned Professor at different Latin American universities and participated as both Trainer and Instructor in projects of the Inter-American Development Bank. He founded—in 2010—the Centre for Study of Sustainable Luxury, first initiative of its kind in the world with an academic/research profile. He is also Founder and Director of the "Award for Sustainable Luxury in Latin America". For his contributions in this field, he was granted the "Sustainable Leadership Award (academic category)," in February 2015 in Mumbai (India). He is Active Member of the UN Global Compact in Argentina.

Dr. Ivan Coste-Manière is Programme Director of Luxury and Fashion Management at SKEMA Business School. His specialties include marketing, communication, and international luxury goods management. He has managed the R&D of several multinational groups and founded several companies in the areas of cosmetics and perfumes, luxury goods, PR and events, and pharmaceuticals. He has been Section Member of the French Economic and Social Council and Deputy Mayor of the city of Grasse for ten years. He has been President, with Prince Albert II of Monaco, of the Track and Field Association: Celebrities for Sports and Charities, while co-chairing the Center for Sustainable Luxury, and works as a consultant for several well-known luxury brands. He holds a Ph.D. in chemistry, is Engineer of the Ecole Centrale Marseille, and has been awarded the Gold Medal (Ministry of Youth and Sports) and the Officer des Palmes Académiques Medal by the Ministry of Research and Education. He is Member of the Economic, Social and Environmental Council Region Sud, of the board of the Comité Français Pierre de Coubertin, Vice President of the Association Francophone des Académies Olympiques, and a distinguished speaker in numerous countries and prestigious institutions and conferences.

Jewellery Between Product and Experience: Luxury in the Twenty-First Century

Alba Cappellieri, Livia Tenuta, and Susanna Testa

Abstract In the contemporary scene, identifying a common and shared definition of what luxury means is ever more difficult and daring. The concept of luxury has changed over time. For centuries, luxury was intended as a sum of beauty and high quality pursued by hand-work and linked to physical products. The digital revolution and the widespread of Information and Communication Technologies has had a significant impact on the global productive system, marking a transition from an 'analogue' era to a digital one—and now 'post-digital'. The paper focuses its investigation in particular on the relationship among jewellery, luxury and sustainability. Firstly, the contribution analyses the concept of luxury and shows how this is increasingly linked to intangible values, where the preciousness of materials has shifted to the preciousness of values. Intangible, far from the needs but close to the desires and dreams of each individual, luxury is seen as the ability to translate the essence of one's time into a product. The examination then focuses on the jewellery field, as one of the most important luxury goods due to its inherent uniqueness and exclusiveness. The entire essay starts from the assumption that giving a univocal definition of jewellery is impossible because of its value and meaning change according to the contexts. Jewellery, as well as luxury, is highly defined by the temporal variable and by the contexts. Today materials are no longer the only characterizing element to define if a jewellery item belongs to the sphere of luxury. Materials and techniques are instead design choices, useful to tell a story. For this reason, the paper takes into consideration examples of high-end jewellery as well as independent brands and designers who use non-traditional materials for their production. The paper identifies as one of the most important value able to represent the contemporaneity the need of awareness. Luxury should tell about excellence, and luxury products stained by the burden of the environmental or social burden cannot be considered as exclusive or desirable. Environmental and social awareness cannot but be part of the production and distribution strategies of companies in the sector. The essay analyses through case studies the different ways for the jewellery sector to promote a sustainable practice. In detail, the first significant grafts for a sustainable supply chain are taken into

A. Cappellieri · L. Tenuta · S. Testa (✉)
Politecnico Di Milano, Milan, Italy
e-mail: susanna.testa@polimi.it

© Springer Nature Singapore Pte Ltd. 2020
M. Á. Gardetti and I. Coste-Manière (eds.), *Sustainable Luxury and Craftsmanship*,
Environmental Footprints and Eco-design of Products and Processes,
https://doi.org/10.1007/978-981-15-3769-1_1

consideration, respecting tradition and the genius loci, the environment and people. Subsequently, the opportunities and risks of progress and technological and digital innovation for the competitiveness of companies in both production processes and lastly communication tools are explored.

Keywords Luxury · Contemporaneity · Fashion Pact · Fashion-sustainability · Jewellery Ethically Minded · CRED · Jewellery

1   Luxury: From Preciousness to Awareness

> Là, tout n'est qu'ordre et beauté,/ Luxe, calme et volupté
> There, all is order and beauty,/ Luxury, peace and pleasure
> Baudelaire, L'invitation au voyage

The concept of luxury has changed over time. For centuries, luxury was intended as a state of great comfort or elegance, especially when involving great expense or an inessential, desirable item which is expensive or difficult to obtain (Oxford Dictionary).

However, this definition of luxury is obsolete or, to be more precise, it is only a mirror of a part of modern society devoted to the convenience and the functionality able to see beauty only in the ephemeral and able to see luxury in what is expensive and opulent. Another part of society, cultured and educated in the knowledge of beauty, recognizes luxury as a life experience, as a result of identity.

The controversy in the definition of what is luxury today already lies in the etymology of the word itself. On the one hand, it refers to the Latin word *lux* that means light. Therefore, something that illuminates, makes visible and understandable, both in terms of logic but also in terms of ethics and aesthetics. So a fully positive meaning, in which luxury is a glorious goal of elevation of status. On the other hand, luxury, from the Latin word *luxus* 'superabundance, excess in the way of life'. Therefore synonymous of superfluous, useless, excess.

These are the two sides of luxury: on the one hand, intuition of beauty and intelligence, and on the other hand, vulgarity and degeneration.

The modern meaning of luxury according to Karaosman [9] dates back to the end of the nineteenth century, when, satisfied with the primary physiological needs—those that Maslow [11] inserted at the base of the pyramid of the hierarchy of needs, luxury was associated with some factors linked to limited stocks, human intervention and value identification. At the beginning of the last century, in addition to functionality, quality, durability and performance, luxury guaranteed a symbolic and experiential value, transcended mere physicality and was characterized by some essential features such as high quality, craftsmanship, exclusivity, uniqueness, provenance, technical performance and the creation of a lifestyle. Some examples are the Mont Blanc pen from 1906 or the Rolex watch from 1908.

Since the mid-twentieth century, technological innovation has triggered changes that have produced economic development and widespread social progress. The third industrial revolution and the economic boom in the 1960s had an impact on the democratization of luxury goods. The democratization of luxury has translated into greater offer and compulsive consumption. This led to the widespread of the so-called Luxury for the Masses [14], and the brands of the sector have expanded their production by inserting more and more 'masstige products'.[1]

Not surprisingly, however, alongside the accessible and widespread luxury, the 1960s are also the years of the first sensational ecological complaints: the American biologist Rachel Carson in her essay Silent Spring [5] showed great concern for the environmental disaster of the last two hundred years, a disaster that will be perpetuated up to the present days. Today's industrial systems, focused on mass production and consumption, have in fact caused an increase in human influence on natural phenomena. Luxury brands offer authentic, excellent and enticing products based on rare and high-quality materials. Yet the raw materials are derived from natural systems constantly influenced by the effects of climate change such as drought, loss of biodiversity and social issues such as lack of skills, loss of employment and health and safety problems.

The changes brought about by industrial revolutions have affected not only the environment, society and consumption trends, but also the concept of jewellery itself [4]. We are witnessing a process products globalization, in which even jewellery, through the reinforcement of the power of the brand, progressively loses its link with the territory in which it is created. The digital shift of the 2000s marked the disruption of industrial production with repercussions on typical territorial districts, with the risk of losing knowledge and know-how linked to traditional craftsmanship.

So, in which direction is luxury going today? And what are the driving factors?

To answer these questions it is worth, first of all, narrowing the field, not talking generically about luxury goods but identifying a specific category. The essay will examine the jewellery field. Throughout history, jewellery has been one of the important luxury goods due to its uniqueness and exclusiveness.

Jewellery, as well as luxury, has over time been the subject of twofold, or rather multiple, interpretations.

Historically, jewellery has always been a land of art, craftsmanship and design. Ambiguous objects with contrasting values, from unbridled luxury to conceptual avant-garde, from the dazzling preciousness of materials to more or less latent design values. On the one hand, art, with the arrogance of its authorship, on the other, fashion, with the transience of its present, in the middle of the jewel with the defence of precious materials as bastions of eternity.

If for a long time the value of a jewel has been synonymous with preciousness, and therefore the physical cost of the material, today this idea is definitely outdated and

---

[1]Masstige is a marketing term meaning downward brand extension, literally 'prestige for the masses'. The term was popularized by Michael Silverstein and Neil Fiske in their book Trading Up and Harvard Business Review article 'Luxury for the Masses'. Masstige products are 'premium but attainable'. These are considered luxury or premium products, and they have price points that fill the gap between mid-market and super premium.

the value of a jewel—as well as an accessory or a product—is the story of everything behind the scenes. It is the result of the quality of the project, the ability of the designer or artist to generate a storytelling around the object thanks to the formal, material or technological choices and production techniques.

Luxury is intangible, far from the needs but close to the desires and dreams of each individual. It is the ability to translate the essence of one's time into a product.

And the materials, the tangible preciousness of the piece, is no longer the only characterizing element to define if an object belongs or not to the sphere of luxury. The material, regardless of whether it is gold, diamonds or wood, is a design choice. Materials and techniques are a means to tell a story. The following chapters will take into consideration examples of high-end jewellery as well as independent brands and designers who use non-traditional materials for their production.

Each object must represent its own time and therein lies its value and meaning. Time is as key an element in the design as manufacture; it moulds the shape of objects, it conditions their function and social utility, it defines the style, the choice of material and technique, it indicates its origins, stratifies its taste and, above all, it reveals its context. Objects are inextricably linked to time. In the case of jewellery, that correspondence is intermittent and discontinuous and, in history, this has not always been able to be the full and mature expression of the Zeitgeist and its climate.

The right question is, then, what is the mirror of our time? What does represent contemporaneity most of all?

The need of awareness. Consumers today, but also producers and people involved in the supply chain, need awareness and aspire to be responsible. What they want is intelligent and responsible beauty, a beauty that exploits tradition but also innovation in favour of thinking processes, materials and strategies. Because the opposite of beauty is not ugliness but ignorance.

And the luxury market has already understood this. It is extremely recent the signing of the Fashion Pact,[2] proposed by François-Henri Pinault, number one of Kering: the largest alliance in fashion, accessories and luxury that concerns the environment and has 32 global companies in the sector. So great that in the full version of the pact, under the heading who says 'we aim for representation of at least 20% of the global fashion industry as measured by volume of products. The goal would be to have a mix of luxury, "mid-level" and "affordable" brands across the fashion sector'. Together they account for about a quarter of the sector [8].

Another recent signal comes from The Union of Concerned Researchers in Fashion that believes that concerned fashion and clothing researchers can no longer remain

---

[2]The G7 meeting was at Biarritz from August 24–26. French President Emmanuel Macron, accompanied by Economy and Finance Minister Bruno Le Maire, Minister of Labour Muriel Pénicaud and Deputy Minister of Ecological and Solidary Transition Brune Poirson, has invited to the Elysée Palace representatives of the 32 fashion and textile companies who have launched the Fashion Pact by his side.

In April 2019, ahead of the G7 meeting, Emmanuel Macron had given François-Henri Pinault, Chairman and Chief Executive Officer of Kering, a mission to bring together the leading players in fashion and textile, with the aim of setting practical objectives for reducing the environmental impact of their industry.

uninvolved or complacent and that researchers need to conduct themselves in new ways. Among their purpose, there is the intention to create an 'activist knowledge ecology', that is, to develop a system of knowledge about fashion sustainability; to advocate for whole systems and paradigm change, beyond current norms and business-as-usual; to express our determined opposition to ill-advised and destructive fashion projects; to take a leadership role in debating existing and new ideas and creating action around fashion-sustainability themes, especially in areas where the generation of new knowledge is of actual or potential significance.

With the aim of spreading sustainability to the production, manufacturing, distribution, consumption and disposal of garments, accessories and footwear, including also social issues, the United Nations Alliance for Sustainable Fashion is operating in that direction. The United Nations Alliance for Sustainable Fashion is an initiative of United Nations agencies and allied organizations designed to contribute to the Sustainable Development Goals through coordinated action in the fashion sector. Specifically, the Alliance works to support coordination between UN bodies working in fashion and promoting projects and policies that ensure that the fashion value chain contributes to the achievement of the Sustainable Development Goals' targets. Fashion, as understood by the Alliance, includes clothing, leather and footwear, made from textiles and related goods.

It is interesting mentioning these examples to understand how different actors—companies, researches, associations and politicians—are facing the issue of being sustainable.

JEM Jewellery Ethically Minded intends to champion an introspective and invested jewellery industry that challenges the status quoin promotes the idea of ethical progress.

JEM is the first French jeweller to be engaged in the «Fairmined» industry. This label guarantees the ethical exploitation of gold, extracted from mines that have subscribed to a process of transformation towards eco-responsible progress. It is by ensuring sustainable development all along the value chain that JEM pursue its mission, in consciousness and transparency. At the same time, JEM commits to a true respect for human beings preserving skilled craftsmanship, particularly French jewellery making.

CRED Jewellery is a jewellery brand involved in sustainability since 1996 and founded to improve the lives of small-scale gold miners. It is immediately clear browsing in its website that opens with a bold message 'CHOOSE CONSCIOUSLY'.

They are a team of people focused the business on a mission to make beautiful jewellery by improving the lives of the communities who mine gold. Jewellery is made using the Fair Trade Gold Standard. From the very beginning, CRED supported sustainability's cause: working in partnership with miners in Columbia to establish the Alliance for Responsible Mining (ARM) to represent small-scale artisanal miners globally and promote socially and environmentally responsible practices; working with mining cooperatives in South America and the Fair Trade Foundation, CRED made the first independently guaranteed fully traceable jewellery in the world to bear the new Fair Trade Gold Mark; using lab-grown diamonds in the UK's first ethical lab-grown ring collection.

About the social impact of jewellery brand, it can be mentioned Eden Diodati, a prêt-á-porter jewellery that works with an extraordinary cooperative of women who survived the genocide in Rwanda. Employing centuries old artisanal heritage and craftsmanship, their skill, courage, fortitude and faith inspires Eden Diodati's creative direction, while challenging preconceptions of 'Made in Africa'. The aim of the brand id shifting paradigms in luxury fashion with a collection of high-end jewellery celebrating the craftsmanship excellence of our partner cooperative.

This makes us understand that the interests of fashion, jewellery and luxury goods, in general, are to succeed in being sustainable—overcoming the so-called paradox of sustainable luxury based on the fact that the production of goods considered 'superfluous' is unethical.

Yet luxury products, including jewellery, are sustainable for three main reasons: they represent a crucial sector for the global economy, as we have just seen from the data of the Fashion Pact; luxury helps to put into practice one of the principles of the ethical consumer, i.e. to buy goods by focusing not on quantity but on quality, thus buying in a more responsible manner. "Buy less, choose well and make it last" is a quota by iconic British designer Vivienne Westwood; true luxury is sustainable and respects people and the environment, taking also in consideration that a brand's reputation increasingly passes through sustainability.

In addition to the striking case of the Fashion Pact, there are other signs that make us understand how the individual jewellery houses have already acquired sustainable logic by proposing models that favour a timeless design, a concept already very close to the dynamics of jewellery.

The concepts of 'timeless design' and 'durability' are, for example, at the heart of Patek Philippe's motto: 'You never actually own a Patek Philippe. You merely look after it for the next generation'.

Many more examples are explored in the following chapters. In detail, the first significant grafts for a sustainable supply chain are taken into consideration, respecting tradition and the *genius loci*,[3] the environment and people. Subsequently, the opportunities and risks of progress and technological and digital innovation for the competitiveness of companies in both production processes and communication tools are also explored.

Going back to the dual etymology of luxury, it seems, hopefully, that the one related to the concept of light is more consistent with the contemporary and that the meaning of luxury is more and more close to beauty, ethics and intelligence than to vulgarity and degeneration professed by the ostentation.

Fortunately, because, as Dostoevskij said in Demons 'Man can live without science, he can live without bread, but without beauty he could no longer live, because there would no longer be anything to do to the world. The whole secret is here, the whole of history is here'.

---

[3]Latin term meaning 'the genius of the place', referring to the presiding deity or spirit. Every place has its own unique qualities, not only in terms of its physical makeup, but of how it is perceived, so it ought to be (but far too often is not) the responsibilities of the architect or landscape designer to be sensitive to those unique qualities, to enhance them rather than to destroy them.

2 A Renewed Sustainable Tradition

Luxury should tell about excellence, the highest outcome of the production process, and luxury products stained by the burden of the environmental or social burden cannot be considered as exclusive or desirable [2]. Environmental and social awareness cannot but be part of the production and distribution strategies of companies in the sector.

There are several options for promoting sustainable models: from the smart use of resources to the reduction of the impact on the environment, from the fair remuneration of workers and the safety of working environments to circular production systems where waste is minimized avoiding to end products in landfills, from safeguarding traditional heritage to fostering the research on new biocompatible materials. In this perspective of process optimization, even in the context of traditional craftsmanship, innovation is essential to reduce the waste of materials and resources.

The goldsmith world has become aware of the need for a reflection on sustainability, a central theme of contemporary thought and future development. In fact, according to some recent research that analyses purchasing behaviour—including 'La sostenibilità cattura Millennials e GenZ' ('Sustainability captures Millennials and GenZ') by Pwc Italia and the Ipsos survey for the 2018 edition of the CSR Show and social innovation—the topic of sustainability is central for young consumers. According to the data, consumers are willing to pay an average of 10% more for guaranteed sustainable and quality products.

The following are some case studies of best practices in the field of jewellery. The analysis takes into consideration both independent designers and established brands that are consciously pursuing the path of sustainability as a positive challenge, able not only to respond to consolidated consumer needs, but also to offer new opportunities for responsible innovation through respect for tradition, human and environmental resources.

2.1 Respecting Diversity and Traditions

In an increasingly standardized global system, cultural diversity is a universal value that must be defended, favoured and preserved (Universal Declaration on Cultural Diversity, Paris, 2 November 2001).

Some realities start from the recovery of the local manual making, of the artisanal practice. Ancient traditions, destined to disappear in the negative effects of globalization, are reviewed in a modern key, allowing not only the revival of techniques and aesthetic models, but also the disclosure of knowledge, conservation and dissemination.

This is the case of Madreforme by Carla Riccoboni. The jewellery collection belongs in all respects to the territory of origin, both in the genesis of the events and in the ability to transfer the values of tradition into contemporary aesthetics.

In 2006, the Angelo Tovo—a Vicenza company whose production was oriented to the creation of moulds for the goldsmith factories of the territory—closed up. Approximately 2500 mother-shapes, the prototypes necessary for the construction of the moulds were recovered and saved by destruction. The jewellery collection, recourse to the use of these moulds, is thus a memory for future generations and a stimulus for new applications of traditional forms in a contemporary way.

Other designers start with the processing of typical traditional materials. Marina and Susanna Sent belong to a family working for generations in the glass sector in Murano, the small island where Venetian glass crafting is based. In 1993, thanks to their know-how and knowledge of production processes, acquired thanks to the experience in the paternal company specialized in glass decoration, they give life to their business. The two sisters, complementary in training, one architect and the other with production-oriented technical studies, propose glass crafted in multiple expressions. They reinterpret the tradition, design the glass and translate it into light modules, transparent bubbles, halfway between pieces of jewellery and garments.

2.2   Respecting People

In this overview of jewellery sustainability, an important chapter is represented by craftsmanship and social inclusion, topics that are often dealt with together. The artisanal practice, in fact, in itself reflects the fundamental criteria of sustainability, giving value to the time of conception and making of a product, to technical experimentation, to the search for quality and perfection and to the passion for what one does. Some productive realities collaborate with local communities, training and offering opportunities for paid work to vulnerable people who have gone through difficult situations on a personal or social level, allowing their redemption. The ultimate goal is to generate business and trigger a virtuous mechanism of independence.

About this, SeeMe is a fair trade verified brand that designs and crafts ethical jewellery. After twenty years of dealing and reporting on women's conditions in the Middle East and North Africa, Caterina Occhio decided to take matters in her own hand creating SeeMe, a training centre and workplace for women otherwise deemed lost. All the jewellery pieces are handcrafted by women survivors of violence. In fact, SeeMe employs women, often single mothers, who have suffered violence and were ostracized from their communities. SeeMe employees learn the craft of jewellery making according to ancient Tunisian techniques. Therefore, while fostering their country's culture and traditions, they secure a workplace for themselves and a future for their families. By wearing the heart-shaped jewellery pieces, people decide to join the Heart Movement, a worldwide movement that wants to replace violence with love. On top of being a safe resort and source of income for women survivors in Tunisia, SeeMe also positively influences established brands in the fashion world, supporting an ethical approach to their sourcing procedures.

Or, again, Indego Africa is a non-profit social enterprise supporting women in Rwanda through education and economic empowerment. Founded in 2007, Indego

Africa partners with female artisans and sells their handcrafted products worldwide. Despite the modern formal appearance, workmanship and materials used for making the jewellery items are those characteristics of the communities in which the artefacts were made. All the proceeds from the sales, together with the donations, are invested by the organization in education programmes for the artisans who make the products.

The work of Riccardo Dalisi, then, combines the theme of training and social inclusion with recycling craftsmanship, providing for the recovery of waste materials to transform them into something new, according to the dynamics of circular economy. Recycling craftsmanship in fact starts from the transformation of materials that would have otherwise reached the end of their life cycle: thanks to design thinking and creativity, the waste materials are transformed into quality products. Riccardo Dalisi was among the first to use poor materials, such as tin, paper, copper, iron, sheet metal, ceramics, glass, wood, fabric, patiently turning them into pieces of jewellery. Self-produced pieces made by hand in his workshop, often with the help of unemployed youngsters to whom he taught a profession, demonstrating that, in a city like Naples afflicted by the waste emergency, jewellery could also have a social value. Dalisi introduced the topic of scrap in jewellery, proving that preciousness is not only that of materials and that tin and paper can have the same dignity as gold and diamonds. For Dalisi, in fact, the difference derives from the need for an ethical vision of jewellery that shuns the preciousness of materials to the benefit of creativity.

2.3 Respecting the Environment

Not only maximizing the social impact, but also minimizing the negative effects on the environment. Pollution in the jewellery sector is often linked to the choice and supply of raw materials.

The extraction of gold, one of the most used materials in jewellery, has over time not only caused the eviction of indigenous communities from the regions concerned, but also polluted the aquifers with chemicals—such as cyanide and mercury—which are used for the extraction of the precious metal from the mineral. Industry is now directing its efforts towards certifying gold meeting international sustainability standards.

Among the forerunners, Chopard, which already in 2013 initiated a programme dedicated to ethical and sustainable jewellery launching the first Green Carpet Collection with Fairmined Gold—that is extracted in certified mines—and diamonds from a producer certified by the Responsible Jewellery Council. The Maison since 2018 uses ethical gold in 100% of the production. Chopard purchases the precious metal exclusively from traceable routes.

The precious stones sector has also taken an active role in promoting sustainability, as shown, for example, by the activities of the Kimberley Process, the certification agreement developed and approved with the joint effort of the governments of many countries, of multinationals diamond producers, and civil society.

Reducing the environmental impact is also possible by creating durable objects and promoting creativity through circular strategies. Producing enough gold for a simple wedding ring creates at least 20 tons of waste. Thus, since 2006, Monique Péan uses only recycled gold and platinum. The designer's recycling activities are not limited to metal, but also include the recovery of precious stones: she travels around the world to find materials that do not require industrial extraction. She collaborates with local artisans and turns to researchers to assess the sustainability of the materials she finds.

Recycled gold, silver and other metals are very on-trend right now as they claim to reduce primary mining and make a real difference to the environment. They are recycled from the Earth's existing metal supply like discarded technology, such as mobiles phones, televisions and computers and make an eco-friendly alternative to the socially and environmentally destructive process of mining the earth.

In this age of technology, with rapid advances and built-in obsolescence, enormous amounts of electronic waste are produced. Furthermore, as gold is a valuable substance worth recycling, it also enables the co-recovery of many other less-valuable metals and ceramics. Some may claim recycled gold is nothing new as the jewellery profession has recycled its metal since artisans first started crafting gold and silver adornment some 5000 years ago and the process for scrapping gold and silver is standard practice within the industry, but still, it is very newsworthy.

The Campana Brothers, the creative duo of Brazilian designer brothers, are part of this scenario of valorizing local materials, often discarded, with the aim of protecting and safeguarding the environmental resources.

The materials, be them humble and modest, or rich and sumptuous, for the Campanas are always the starting point for the design process, from chairs to lamps, from fashion accessories to jewellery. Bones' Structure, for example, is a modular necklace formed by a set of scrap leather parts that can be assembled in various ways by the consumer through magnets. Production waste usually has irregularities that make the pieces unique and never obvious in their imperfection.

The concept of modular product is itself part of global sustainability perspectives because it offers the possibility of combining the elements in countless configurations, satisfying the need given by different occasions of use. In fact, modular jewels allow the user to intervene on the arrangement of predefined parts, to combine them and wear them in various ways. This is the case of Precious Molecules by Massimiliano Adami, a collection made up of submultiples, assembled parts, each of which forms the core of the structural elements of the jewellery piece. Units to be assembled as parts of a kit, as in the Drilling Lab Project which creates a precious alphabet reinterpreting the structures of industrial clamps; Manuganda's Compo describes a modular chain that can be disassembled and reassembled in different ways, stimulating the creativity of the wearer.

With a view to global sustainability, research is moving towards the development of ecological and performing materials and the optimization of production processes in order to reduce resource consumption and the use of toxic substances. The key to renewal lies in interdisciplinary experimentation, the outcome of a mixture of sectors, skills and processes, in which science, ethics and aesthetics are integrated to create something new.

Today experiments on alternative non-polluting materials are increasingly frequent. Taking into consideration diamonds, one of several issues that the industry is buzzing about is lab-grown ones. Consumers are asking for them even if there are not on the market yet. The appeal of lab-grown diamonds goes beyond their being priced 20–30% cheaper at retail than mined diamonds. Consumers are drawn to them because they do not contribute to the destruction of the environment and no one is harmed in their creation, two big considerations that tie into the 'responsible sourcing' movement that is gaining traction among consumers, particularly Millennials. Therefore, manufacturers should surely begin thinking about incorporating them into their lines.

These options have a number of benefits as they require less human labour, have a reduced carbon footprint, are indistinguishable from their natural counterparts and come with a significantly lower tag price. But at the same time, designers warn against using artificial stones exclusively, since could deprive entire mining communities of their livelihood.

The case studies shown above are just some emblematic examples of a global vision that aims to combine aesthetics and technique with an ethic that is attentive to the choice of materials, to the respect for the environment, to the recovery of the collective heritage, to the know-how linked to the territory.

The new work system provides for the sharing and optimization of resources in a relationship open to change. About Fucina Orafa, for example, is an Italian reality, a place where it is possible to create and learn, with a view to sharing knowledge, saving resources and spaces. It is an independent laboratory for goldsmiths, students and artists linked to the world of art and jewellery, a network of professionals who collaborate with each other, keeping alive the goldsmith traditions combined with contemporary techniques in order to achieve high-quality jewellery pieces.

3   The Digital Shift: The Impact of Digital Technologies on Jewellery

The shift from a handcraftsmanship-based methodology to new models of productive processes was marked by the first industrial revolution: the handcraft know-how had been gradually flanked and, for some sectors, entirely replaced by the use of machine tools, with significant benefits in terms of quantity, speed and efficiency. It is during the industrial revolution, the transition from the artisanal manufacture to the mass production, that for the first time the contrast between *manus* (hand) and *machina* (machine) comes to light, two opposing elements which have characterized every productive and artistic sphere for the centuries ahead [1].

Also, the jewellery sector, thus, has been affected by the dichotomy generated from the manufacturing revolution.

Hand manufacturing and mechanized production, respectively, hand and machine, have marked and settled over time two opposite ways of creating and therefore conceiving jewellery: from the one side the manufactured object, aimed at an elite and associated with luxury; on the other side the piece of jewellery industrially produced, addressed to the mass market.

The current revolution has marked the transition has globally fluidified the system: in the age of digital technology, the distinction among the different spheres is increasingly less defined, the productive paradigms tend to blend and to contaminate each other, reducing progressively the antithetical distance between hand and machine. Even though the dichotomy is still representing a significant characterizing element for the jewellery field, the boundaries between the two different production processes are becoming more and more blurred and the cases of a hybrid approach are more frequent [18].

The artisanal practice of making jewellery is in fact more often integrated with the use of digital technologies, such as the one of additive manufacturing, and the creative processes are shared online. The precision and careful attention to details that has characterized traditional jewellery making can be flanked by a new dimension, that is the one of the web, considered not only as a network of processes and data but as an integrated and interactive system that holds together people and things. In this context, small- and micro-companies, using new flexible production technologies, are able to produce small batches or even unique pieces, customized with different finishes, so as to be able to combine artisanal quality with industrial processes.

To this end, some interesting case studies are Bijouets and Maison 203, two of the main Italian companies that have introduced 3D printing technologies into jewellery without neglecting craftsmanship quality and formal beauty. All the 3D printed pieces are rigorously finished by hand and conceived by professionals coming from other sectors, such as architecture and product design, who contribute in terms of cross-fertilization as well as contemporary aesthetics. The result is a rich and varied production that intersects design, technological experimentation and craftsmanship [3].

Bijouets, founded in Trento by Ignazio Pomini and Fabio Ciciani in 2010, is dedicated to making accessories with professional 3D printing techniques. Pomini immediately understands the potential of this jewellery technology and, with the creative direction of Selvaggia Armani and with the contribution of different designers such as Federico Angi, Filippo Mambretti or Maria Jennifer Carew, Bijouets becomes a reference for 3D jewellery at an international level.

Another Italian company of reference is Maison 203, which represents one of the most interesting examples of digital craftsmanship of accessories in nylon and PLA (a bioplastic derived from corn starch), to which are added special editions in steel and polished brass. The company was founded in Treviso in 2011 by designers Orlando Fernandez Flores and Lucia De Conti and design is one of the fundamental ingredients of their production, which relies on the collaboration of the best Italian designers such as Odoardo Fioravanti, Giorgio Biscaro and Giulio Iacchetti.

If the jewellery of Bijouets and of Maison 203 are declined according to the languages of fashion, other designers choose 3D printing to make artistic jewellery pieces, as happens with Ross Lovegrove who created 3D printed gold items for the

London gallery of Louisa Guinness, demonstrating that this technology also fits the needs of art.

The introduction and diffusion of digital communication and production technologies—has entailed a radical change which has affected the very notion of artefact, luxury and sustainability. Aesthetics and technical skills have now been joined by the ethical selection of materials, respect for the environment and the protection and promotion of the joint cultural and traditional heritage associated with the surrounding territory. Biology, chemistry and engineering have greatly contributed to the identification of solutions favouring sustainability, thanks to the development of new materials that are either recycled or made of biocompatible and biodegradable natural fibres, as seen with the pieces made in 3D printed PLA by Maison 203; digital technologies, however, can also supply some solutions which may make the production more sustainable.

Indeed, digital technologies have favoured the transition from centralized and large-scale systems of design and production to decentralized or distributed systems in smaller units and connected to the network [17]. The production distributed on demand reduces the resources used, the costs of production, packaging, transportation and even the waste of finished products that remain unsold. The artefacts are therefore territorialized, and they are produced closer to the final consumer, thus reducing the pollution linked to the use of superfluous resources and to transportation.

Modern design arose from industrial chain production with great output and brought about the aesthetic standardisation of the industrial offer, whose only lingering value resides in its very brand name; design's widespread presence and distributed production, however, Srai et al. [15] do not supply only products, but also services which are instrumental in the creation of customized objects, and which actively engage consumers by educating them and stimulating their creativity. The serial production of the finished product is replaced by the design of models, digital platforms and semi-finished products which may be modified, customized and finished by the consumers themselves.

The attribution of an active role to the consumer, his involvement in the practical definition of the artefact, makes him more sensitive to the problem of waste. Technology has led to a massive penetration of design into the everyday life of consumers, who have progressively taken on a much more active and critical, more attentive and responsible role and claim the right to be able to take part in the creative phase [7].

The dynamics surrounding the cooperation and web-based design stage is somehow steeped in an 'achronic dimension' [13], in which the various stages may overlap: the implementation and creation stages take place simultaneously, and it is indeed possible to consider also the design, production and distribution stages as synchronous. The design aspect expands and moves from a closed to an open condition and consequently evolves from being a personal expression of individual talent to a collective profession [3]. The 'open' qualification applied to digital design needs to be contrasted with industrial design, whose output is by definition closed: in serial production, the object is defined conclusively, and the creator and the interpreter coincide in the same person [7].

These phenomena seal the final abandonment of the serial production paradigm. Industrial production traditionally envisaged the exact reproduction of a model as a series of identical objects; over the past few years, technological development and social change have determined the elimination of the contraposition between unique pieces and reproduced ones, between the model and its reproductions, and production approaches have become blurred. This has been made possible by the creation of automatically finished pieces from a drawing; such a change emerged gradually in the last decades of the twentieth century and has entailed renouncing the idea that a project is definitively finished in itself.

The coupling of the web with the new technologies permits the elaboration of a collective project in which the final product is determined by the buyer: a sort of hybrid object suspended between industrial production and individual creation. It has therefore been a veritable revolution which has changed concepts such as serial production, multiples, processes, as well as the role of designers, producers and consumers; the final result cannot be anticipated.

It is not just consumers who can now select and create their own objects industrially: digital technologies have also been able to provide diversified solutions throughout the production chain in answer to the demand for tailor-made products, services and experiences which may be endlessly customized.

The real revolution which the development of new technologies has entailed thus consists in the possibility to autonomously create objects which totally match individual needs, desires and size. The contemporary world is witnessing a changeover from standardized production to one that is more flexible and oriented towards consumers' customization: objects are designed ad hoc and are produced in a system that conjugates the potential of mass production plants with the flexibility afforded by the new hardware and software.

The widespread presence of digital technologies has optimized and concretely improved jewellery companies' performance, but has also provided the possibility to affect the final users' experience. Indeed, the new powerful technological tools available constitute an open system and manage to involve consumers at all stages.

This systemic interaction generates an added value in terms of experience and amplifies the emotional component associated with the product.

Indeed, the new technologies enable designers to dynamically engage final users online already during the designing and production stages, in order to achieve a product which may fulfil the customers' needs as much as possible both in terms of wearability and aesthetics.

An example of this is the Orchid by Studio Minale Maeda for Chi Ha Paura...? The project consists of a 3D printed orchid boutonniere that can attach to a buttonhole as well as be clipped on any garment by its own stem. The shape of the flower is generated each time anew when it is bought, based on a variety of influences like location, climate, soil, season and many more. If one does not like the result, one can wait for different conditions to get a different orchid before purchasing it.

Digital co-creation platforms enable the establishment of a new virtual space for cooperation, in which designers no longer define the object's final structure, but in fact regulate and programme the procedure itself: designing therefore consists in defining algorithms on the basis of parametric dynamics which can generate structures with endless variants while retaining a consistent overall matrix. A pioneering example of virtual interaction in the jewellery sector is supplied by the nervous system, which thanks to generative systems, 3D printing and digital platforms releases online apps that make design more accessible by allowing users to create their own products. These tools empower users to interact directly and intuitively with the graphic interface; users can thus create endless formal variations and generate structures inspired by nature or by the fractals' perfect geometry, complex and ever-changing figures which constitute unique pieces and which may be booked for ad hoc production.

Virtual platforms have also emerged which make the production chain shorter and more interactive, still with the final goal of letting customers play a greater role in the process. I materialize, Shapeways, Sculpteo and Thingiverse are only some examples among the many which let users upload their own digital design, select their material of choice, prototype it and possibly even sell it online.

The new technologies go beyond customization and unique pieces and even go as far as tailor-made production. Three-dimensional scanners can measure and trace precisely and rapidly the three-dimensional shape of the human body on which the product will be modelled.

Three-dimensional scanning, the new frontier of sizing, constitutes a veritable virtual leap forward. It is possible to create tailor-made items simply by scanning the body; the scan will form the basis for a three-dimensional model, which once worn by an avatar will provide an early idea of the physical prototype. Thanks to virtual prototyping these models may be consulted, tested, modified and rectified just as if they were physical objects, without having to resort to any production resources or materials.

The possibilities offered by 3D scanning technologies are not confined to process optimization or to the goal of providing end users with a product that perfectly fits their body beyond the notion of 'size', but actually extent to suggesting interesting starting points for artistic and designing experimentations. Suffice it to consider Portrait me, Vivian Meller and Laura Alvarado's project that reinterpreted in a modern and ironic perspective the traditional cameo: the subjects were dressed in re-enactment clothing, faithfully reproduced by the 3D scanner, and the acquired data was then printed in a collection of brooches by means of selective laser sintering.

4 The Power of Communication: From Exclusivity to Inclusion

The Latin *caveat emptor* means 'let the buyer beware!'. It is composed of the verb *caveo*, used in exhortative and impersonal form, and the noun *emptor*, buyer.

Born at a time when there was no legislative protection for economic actors, this way of saying has remained relevant over the centuries. Although jurisprudential law, through the discipline of unfair competition, has issued rules prohibiting dishonest behaviours aimed at deceiving the public, contrary to the principles of good faith, loyalty and honesty, it is a good practice still today to pay attention when buying any good or service. Specifically, greater attention should be paid during the most particular and least frequent occasions of purchase, that is, when one buys a good that is very expensive or that is not part of daily consumer behaviour.

This expression is still very relevant in its ability to highlight the importance of information for the consumer.

Today, consumers must be aware of what they are buying and, at a time when obtaining the information they are looking for has become very fast and easy, it is not justifiable to behave in a disinterested manner. However, this excess of information—from the print media to social media, radio, television and blogs—requires greater consumers' participation in selecting contents.

In the early years—more or less throughout the first half of the twentieth century—of mass communication the media and messages available were limited number. There were very few sources of communication and a symbolic universe not as crowded with messages as today. Over the years, especially thanks to technological innovation, the number of broadcasters and the number of messages has increased dramatically, so that today we can precisely speak of a situation of abundance.

This not only generates, as already mentioned, an orientation effort in the person who receives the message but also in the person who sends it. The downside is, in fact, that in a mare magnum of information, brands or producers must ensure that they find a way to be heard, to create a priority between the information they transmit and communicate effectively to the public.

Communicating means, in fact, first of all, to transmit a message to one or more people. A subject, action or message begins to live as soon as it is communicated. Secondly, communication is the elaboration and sharing of meanings, which means organizing a structured and understandable communication that does not frustrate all the efforts of companies that operate in the name of sustainability. Ultimately, it means building and modifying relationships, creating a community, but also making sure that there is a responsible consumer who listens and is able to make a critical and ethical assessment.

Communication is structured like a chain, and it only takes one link to weaken to make it less effective or completely powerless.

The first link, the key element of the message, is the issuer who, through a channel of transmission, sends the body of the conversation, the message, which in turn contains a referent, or topic and is structured according to a code. Finally, through a reception channel, the receiver receives the message sent by the issuer.

In this chain, each character has its own responsibility. The theme of sustainability, therefore the message, is extremely complex but at the same time it is a priority. Therefore, in order to ensure that the message arrives unhindered from the issuer to the receiver and is absorbed correctly, four factors are needed that, in communication theories, guarantee the effectiveness of the result: listening; mutual interest; understanding and acceptance; and changing an attitude.

4.1 Listening

Listening means creating the conditions for the issuer to send out a message and make sure someone gets it. A statement by Manuel Castells, now famous, compares today's speed with the previous rhythm of change: 'in the United States the radio has taken thirty years to reach sixty million people, television has reached this level of diffusion in fifteen years, the internet has done so in just three years since the birth of the world wide web'.

Today, reaching a very wide audience has become extremely simple and democratic. Luxury, which has always been undemocratic by definition, has also succumbed to the new policies generated by digital communications. Particular reference is made to social media, one of the most popular media today.

Many of the jewellery brands have in fact approached digital marketing strategies. As an example, David Yurman for several years has been collaborating with the most popular Instagram influencers—Wendy's Look book, Brooklyn Blonde and Atlantic-Pacific, uses Instagram's social shopping function to increase sales directly from within the app and invests in collaborations such as the one in 2017 with Elle for the first Facebook live stream of the magazine. The partnership not only allowed them to reach Elle's audience of 4.7 million Facebook followers in addition to their own, but also drew on the power of two influencer guests of Instagram, Erica Hoida and Lucy Hernandez, who at the time had a total of over 880,000 followers.

Bulgari, one of the most famous high-end jewellery brand, has also been successfully implementing digital jewellery campaigns with the support of influencers for some time now. This year the brand worked with four major Arab influencers—actress Tara Emad, model Rym Saidi and style influencers Lama Al Akeel and Fatma Husam—in a dazzling campaign directed at the Arab market. Bulgari shows particular attention in conversations generated by hashtags that encourage followers to create personal content.

Buccellati, the famous Milanese fine jewellery brand, is also involved in digital strategies. One of the most interesting recent campaigns has been the collaboration with Noonoouri, a virtual influencer with over 86,000 Instagram followers.

It is clear that the role of mediators, in this case of influencers, is a fundamental tool to facilitate and amplify communication channels. The use of influential characters does not only occur, however, on social networks, but also in the actual collaboration on the product.

In particular, in the field of sustainability, the collaboration between Penélope Cruz, a Hollywood star and very committed humanitarian activist, and Swarovski is exemplary. They are working together to pave the way for the creation of sustainable jewellery, without giving up on luxury design. In June 2019, they launched the latest collection of conscious luxury jewellery, using only responsibly sourced materials.

One of the goals of the collaboration is to encourage jewellery buyers to be more aware of the benefits of environmentally friendly practices and inspire the individual to make a difference. This is the first time Swarovski has worked with a major celebrity to create an ethical jewellery collection, and they follow in the footsteps of Chopard, which has co-created designs with the likes of Marion Cotillard and Cate Blanchett.

At the same time, more and more platforms are being created, both real and digital, dedicated specifically to the theme of sustainability and its intertwining with the worlds of luxury. An example can be Eluxe Magazine, the world's first-ever publication fully dedicated to sustainable luxury. It is a quarterly published paper magazine and a daily updated digital publication based in London, dedicated to showcasing luxury brands that demonstrate a strong commitment to good ethics and environmental sustainability.

If the examples cited so far, however, see as the final user mainly the consumer, there are, on the other hand, also 'listening areas' for professionals. In these areas, all the activities involving the actors of the supply chain are presented. One example is the Responsible Sourcing Blue Book produced by CIBJO, which provides a framework and guidance for ethically sourcing gems and precious metals responsibly in the jewellery sector. Or the Responsible Jewellery Council helps companies of all sizes, throughout the jewellery supply chain, meet the rising ethical demands of peers, consumers, financial institutions and civil society.

It does so by providing a clear set of standards—the RJC 'Code of Practices'—which is verified through a third party, independent, certification process. Adoption and adherence with the RJC's Code of Practices presents a pathway for companies to address sustainability best practices and align with the 17 United Nations Sustainable Development Goals.

4.2 Mutual Interest

The second phase, after listening, is the creation of a conversation. The actors involved in the communication are both active parts. Because of the leading role, social media have nowadays luxury fashion consumers which require an interactive approach from brands. This is high contrast with the past single-sided selling strategy, needs and demands of consumers are becoming a pillar on which to build the approach to the market. On the basis of that, social media is thought to be a great way for luxury brands to gain information about market needs from customers themselves [12]. Social media created the necessity for luxury fashion brands to move from very traditional advertising strategies to SMM (social media marketing). This latter approach is more of slow-building, delivering solutions, generating curiosity and seeking feedbacks for a continuous improvement of goods, services and customer relationships. Indeed, social media can act as a great deliverer of the luxury dream, a pillar of the whole industry. This because through platforms like Instagram, Facebook, Snapchat and so on companies can communicate a lot with potential customers and fans, utilizing all the tools and features of those websites.

The conversation about the theme of sustainability in jewellery is now very heated. Just open Instagram and look for the hashtag #sustainable jewellery to see that it appears in 75.8 K post, the hashtag #ethical jewellery in 120 K post, #recycled gold in 44.9 k post. Generally, hashtags were born with a specific purpose: to involve web users on a particular topic and group all these conversations under a single hat in order to generate new conversations or participate in the existing ones. The power of hashtags is that they instantly expand the scope of messages, intercepting, in addition to followers, also all those users who are not followers but who are interested in that particular topic.

For this reason, hashtags are an excellent gauge to understand how much the public is involved in some issues and what are the trends.

Looking for the #sustainable jewellery hashtag, for example, immediately suggests new related themes such as recycled jewellery, ethical jewellery, recycled gold, eco-friendly jewellery or artisan jewellery.

However, in addition to the digital worlds, a series of places and events have also been done with the aim of creating a network around the theme of sustainability in jewellery but also in an attempt to disseminate the processes already underway.

An example is the Andrea Palladio International Jewellery Award, an international award dedicated to the excellence of contemporary jewellery in the fields of design, production, retail and communication and promoted by Italian Exhibition Group. Among the various categories awarded, the 'JEWELLERY CORPORATE SOCIAL RESPONSIBILITY AWARD' stands out since 2014. It refers to the best goldsmith company that has based its production processes on responsible practices from an ethical, social and environmental point of view, with respect to human rights, from the extraction of precious materials to the marketing of the finished product.

Since the first edition in 2014, personalities of the calibre of Eli Izhakoff have been awarded for their activity in the world of jewellery from an ethical, social and environmental point of view, Caterina Occhio for the SeeMe project and brands such as Pandora, which has made social responsibility one of the constraints on which to base its decisions in terms of design and materials, or Chopard for its commitment to the project with Eco-Age.

This type of initiative helps to strengthen the dialogue between the various players in the supply chain, as well as involving new ones. Furthermore, the establishment of an award helps and generates and gives visibility to models that can act as a driving force for the realities of the same world.

More and more frequent are also the events and conferences organized around the specific theme of the sustainability of jewellery. One example is the Chicago Responsible Jewellery Fair, founded in 2017, when jewellery designer Susan Wheeler decided to bring people together across the world to discuss how to make jewellery supply chains more transparent and make jewellery business more beneficial to all members of the industry vertical. The CRJC's mission is to engage everyone in the jewellery industry: miners, makers, professionals, educators and students. To address all the ways that individuals and companies can be involved in the responsible jewellery movement. To make a difference by making things happen.

Or also the Jewellery Industry Summits—the first in 2016—invite a broad group of designers, manufacturers and retailers to share their perspectives, experiences and ideas for a better way. This inclusive approach creates an opportunity for participants to take steps to improve responsibility and ethics in the creation of jewellery, no matter where they fall in the supply chain or what stage their business is at.

Vicenzaoro also dedicates lots of seminars to the topic of sustainability. At Vicenzaoro September 2019, the topic of the seminar organized by CIBJO, The World Jewellery Confederation, is sustainability and responsible sourcing in the Jewellery industry.

4.3 Understanding and Acceptance

Once the message has been generated, listened to and has aroused interest in the receiver from the next phase is that of the acquisition.

Specifically, awareness is generated in the buyer who becomes able to make a critical and ethical assessment. The consumer begins a 'sustainable thinking' having understood the importance of the traceability of the supply chain, the transparency of processes and materials, optimization of water, energy and chemical consumption.

4.4 Changing an Attitude

Lastly, the information is understood and acquired to such an extent that it had a real effect on the consumer and consequently on the purchase.

At the time of purchase, therefore, the consumer wonders about the life cycle of the product and must be able to receive all the information he or she needs to make a responsible choice. In this, many brands are acting in an exemplary way to make each product transparent and traceable. If the communication strategy in the past was aimed at transmitting information about the physical features of the jewellery piece, today it is storytelling-in-action focused, conferring more importance to the immaterial and experiential values and also to the 'life' of the product itself.

The product itself, through digital labels, QR codes, for example, can tell their story, and this clearly represents a competitive advantage for the company.

Being transparent means acting in the sign of sustainability and not being afraid to tell how to act, but, on the contrary, trying to communicate it as much as possible. Companies striving towards sustainability are not afraid to report the targets they have set and transparently show how far they have come and how much further they need to go. At a minimum, material issues to report include supply chain, traceability, fair trade, waste management and environmental pollution. Large companies such as Pandora and Tiffany & Co have started such reporting.

Tiffany & Co. has a reputable mine-to-market programme that provides consumers with the transparency and traceability through its Social Accountability Programme, which traces the company's products as they evolve throughout the supply chain.

Also, the Diamond Source Initiative is the most recent launched by Tiffany & Co., and led by Anisa Kamadoli Costa, chief sustainability officer of Tiffany and senior vice president diamond and jewellery supply Andy Hart, in early 2019 to promote traceability and transparency in the diamond industry. Consumers will be able to know the origin (region or country of origin) of all diamonds recently mined above 0.18 carats. Each diamond is engraved with a serial number and customers receive a certificate of origin for the diamonds. By 2020, the New York giant will also be able to share information on the intermediate stages, including where diamonds are cut and cleaned.

For this initiative, Tiffany & Co. uses a proprietary and secures database that links the serial number of the diamond to its origin and craftsmanship. Yet the traceability of diamonds is mainly associated with the blockchain technology that is still evolving today. When the Kimberley Scheme for conflict-free diamonds was introduced in 2003, it was one of the earliest traceability programmes in ethical sourcing. But today blockchain technology is coming to the jewellery industry to guarantee absolute traceability of every element in the jewellery supply chain. Blockchain tracks raw materials—like gold or diamonds—from the mine to the refiner or gem cutter, through distributors and manufacturers, to the retailer and to the consumer. But the real innovation it is that instead of keeping a paper log, the raw material is assigned a serial number, and data is entered into the digital ledger, as it moves from place to place, throughout the material's transformation. Blockchain systems cannot be edited and they are extremely secure, so the data that is entered for each step of the process is permanent. The technology enables diamond suppliers to replace a paper certification process with a blockchain ledger.

Some companies are applying blockchain in their supply chain tools. Everledger Diamond is a traceability initiative built on a blockchain-based platform for the diamond and jewellery industry with the aim to engage all industry participants including manufacturers, retailers, and consumers to know a diamond's story from the origin to the end customer.

Trustchain is a blockchain created by IBM that proves the provenance of jewellery by following the supply chain from mine to store. It includes a consortium of companies involved in every step of the supply chain: Asahi Refining, the precious metals refiner; Helzberg Diamonds, a US jewellery retailer; LeachGarner, a precious metals supplier; and The Richline Group, a global jewellery manufacturer.

Tracer is an online distributed ledger based on blockchain created by DeBeers that improve global provenance in the diamond and jewellery supply chain. It is very significant information because 70% of the world diamond comes from DeBeers sites.

Bit Carat is a start-up with the aim to create an asset-backed token that allows the trade of certified and safe diamonds through traceability using the blockchain, ensuring the origin of natural diamonds throughout their history.

Currently, on the market one can find natural diamonds, fake diamonds and synthetic diamonds. Synthetic diamonds have the same physical and molecular characteristics as natural diamonds, but they are human-made and do not take 3 billion years to form like natural diamonds. The process for their production is costly due to high-energy consumption, but it has been now so refined to produce diamonds which are difficult to distinguish from natural. However, synthetic diamonds have a value of 40% lower than the natural ones.

In order to be able to distinguish the two types of precious stone, the blockchain could be the ideal solution. This because the blockchain can create a unique token for each natural diamond, with the physical characteristics of the product that cannot be changed, and that can be recorded and traced across the different steps; from the mine, to the cut to the final buyer. Also, certified security tokens would allow easy transfers among buyers.

The evolution of blockchain technology reveals optimistic scenarios for consumer information. It is not difficult to imagine how, in a few years, the consumer could be able to use a smartphone to determine a gem's provenance.

5 Conclusions

The essay highlights the contemporary challenges at the intersections of three main areas: luxury, jewellery and sustainability.

It does this by analysing the opportunities and limits of the enhancement of goldsmith artisanal tradition but also taking into consideration the opportunities offered by digital innovation. It is important to underline that one does not provide for the exclusion of the other, but rather as analogue and digital, hand and technology find their greatest balance and their maximum enhancement in coexistence.

The very rise of a new kind of consumer, more critic, aware and responsible, with 'sustainable thinking', makes it possible to dare with more fluid assessments that do not include categorical judgments in favour of one or the other approach, understanding the life cycle of the product at 360 degrees. Precisely for this reason, it is necessary to communicate information, to do it clearly in order to involve and convince the consumer to invest in sustainability.

There is a famous phrase from the science fiction writer William Gibson that says 'The future is already here—it's just not very evenly distributed'.[4] It is something to be built together, creating a dialogue and fostering harmony between designers, companies and consumers and, based on their and our ability, transforming everyday life.

Contributions

1. Luxury: From Preciousness to Awareness by Prof. Alba Cappellieri.
2. A Renewed Sustainable TRADITION by Dr. Susanna Testa.
3. The Digital Shift: The Impact of Digital Technologies on Jewellery by Dr. Susanna Testa.

---

[4] 1990, Cyberpunk (Documentary), Directed by Marianne Trench, Produced by Peter von Brandenburg, An Intercon Production. [Excerpt occurs in Part 3 of 5 parts; Timecode 12:20 of 14:59] (Video available in 5 parts on youtube; Viewed on 2012 Janaury 24) link.

4. The Power of Communication: From Exclusivity to Inclusion by Prof. Livia Tenuta.
2. Conclusions by Prof. Livia Tenuta.

Bibliography

1. Bolton A (2016) Manus x Machina: fashion in an age of technology. The Metropolitan Museum of Art, New York
2. Brun A (2019) Ecco perché il lusso, quello vero, è una cosa buona e sostenibile. ilsole24ore.com. Availabe at: https://www.ilsole24ore.com/art/ecco-perche-lusso-quello-vero-e-cosa-buona-e-sostenibile-ABpoXcmB?refresh_ce=1
3. Cappellieri A (2016) Brilliant!. The Futures of Italian Jewellery, Corraini Edizioni, Mantova
4. Cappellieri A, Tenuta L, Uğur Yavuz S (2017) The role of design for the brand identity of jewellery. In: Thiene WM (ed) Luxusmarken-management. Springer Gabler, Germany
5. Carson R (1962) Silent Spring. Houghton Mifflin, Boston, Massachusetts, USA
6. Castells M (2000) The rise of the network society. Blackwell Pub. Blackwell Publishers, Inc. Cambridge, MA
7. Dubois D (2017) Digital and social strategies for luxury brands. Springer, Fachmedien, Wiesbaden, Germany. https://doi.org/10.1007/978-3-658-09072-2
8. Finessi B (2014) Il design italiano oltre la crisi. Autarchia, austerità, autonomia. Catalogo della mostra, VII Edizione della Triennale di Milano. Corraini Edizioni, Mantova
9. Guilbault L, Kent S (2019) Kering Chief to Present Industry Sustainability Pact to G7. businessoffashion.com. Available at: https://www.businessoffashion.com/articles/news-analysis/kering-chief-to-present-industry-sustainability-pact-to-g7
10. Karaosman H (2019) La moda di lusso deve promuovere una gestione etica e ambientale. In: Bettoni G, Natalini AS, Ricci S, Sozzani Maino S, Spadafora M (eds) Sustainable thinking. Mondadori Electa, Florence, Italy
11. Maslow A (1954) Motivation and personality. Harper & Brothers, New York, NY
12. Okonkwo U (2009) Sustaining the luxury brand on the Internet. Journal of Brand Management. 16:302–310. https://doi.org/10.1057/bm.2009.2
13. Rubinstein H, Griffiths C (2001) Branding matters more on the Internet. J Brand Manag 8:394–404. https://doi.org/10.1057/palgrave.bm.2540039
14. Sbordone MA (2012) Discronie. Fenomeni del contemporaneo nella Moda e nel Design, Alinea Editrice, Firenze, Italy
15. Silverstein MJ, Fiske N (2003) Luxury for the masses. Harvard Bus Rev 81(4):48–57, 121
16. Srai JS et al (2016) Distributed manufacturing: scope, challenges and opportunities. Int J Prod Res 54(23):6917–6935. ISSN 0020-7543
17. Tenuta L (2020) La moda nell'era digitale. Nuovi prodotti, nuovi processi e nuovi servizi / New Products, New Processes and New Services. Aracne ISBN 978-88-255-3112-1
18. Testa S (2019) FashionTech. Body Equipment, Digital Technologies and Interaction. Universitas Studiorum, Mantua (MN), Italy ISBN 978-88-3369-057-5

Sustainable Luxury, Craftsmanship and Vicuna Poncho

Roxana Amarilla, Miguel Ángel Gardetti, and Marisa Gabriel

Abstract Sustainable luxury presents as a meeting point for traditions, cultures, values and needs. It offers an opportunity to rescue and expand the cultural heritage of communities, enhancing their history to share it with the world. Craftsmanship is at the heart of sustainable luxury because artisans are the ones who add value to objects. The Vicuña Poncho produced at Laguna Blanca, Catamarca, is an object with a deep symbolic character which is representative of such community. Artisans show their sense of belonging, unity and values in this garment. As they continue with these practices, they can teach the most exquisite technique and reflect the ancient knowledge of their ancestors—the indigenous communities of the region. Meanwhile, they create a top-quality finished product that can be framed into sustainable luxury. The following chapter deals with sustainable luxury, craftsmanship and handcrafted quality. The chapter introduces some concepts that offer a theoretical framework to subsequently delve into the Vicuña Poncho universe. It offers a description of both this object, highlighting its historic and current symbology, and the ancient aboriginal practices still used by artisans, to conclude with thoughts about the importance of sustainable luxury and the appreciation of the Vicuña Poncho to empower artisans and expand their horizons.

Keywords Vicuña · Poncho · Sustainable luxury · Craftmanship · Laguna Blanca · Catamarca

R. Amarilla (✉)
World Crafts Council Latin-American, Carlos Pellegrini, 485, 1st Floor E, Buenos Aires, Argentina
e-mail: roxanaamarilla@gmail.com

M. Á. Gardetti
Center for Studies on Sustainable Luxury, 5th "B", Paroissien, 2680, C1429CXP Buenos Aires, Argentina
e-mail: mag@lujosustentable.org
URL: http://www.lujosustentable.org/

M. Gabriel
Sustainable Textile Center, 5th "B", Paroissien, 2680, C1429CXP Buenos Aires, Argentina
e-mail: mara@ctextilsustentable.org.ar
URL: http://www.ctextilsustentable.org.ar

© Springer Nature Singapore Pte Ltd. 2020
M. Á. Gardetti and I. Coste-Manière (eds.), *Sustainable Luxury and Craftsmanship*, Environmental Footprints and Eco-design of Products and Processes, https://doi.org/10.1007/978-981-15-3769-1_2

1 Sustainable Luxury, Craftsmanship and Handcrafted Quality

1.1 Concepts and Thoughts

Within the framework of globalisation, both fashion and luxury are impoverishing cultural diversity and promoting the consumption of unsustainable clothes and objects. Therefore, the wealth and local production of each region is replaced by goods from other countries, which lack cultural representation. "Economic and cultural globalization, through its mechanisms of levelling life, consumption, and product recreation, has led over the past 25 years, to the destruction of national identity in products to an overestimation of international brands" [17: 101].

Historically, luxury has been characterised by offering noble materials, celebrating the handcrafted quality and cultural heritage to produce unique, timeless, long-lasting objects. Given their great symbolic value, luxury objects were usually passed down from generation to generation. Sustainable luxury attempts to go back to the roots of luxury. It means going back to the thought over purchase, craftsmanship and noble materials, respecting both the environment and the society. Consequently, it offers an opportunity to recover, preserve and promote the cultural heritage of different regions and communities.

While, at first, this can pose a paradox, sustainable luxury has the potential to protect species and ancient traditions, as well as to enrich the diversity and cultural heritage. "Sustainable luxury appears as a connective environment, where needs, values, and cultures are collectively shared" [10: 16]. It is based on the creations of artisans and masters, in order to spread knowledge which, otherwise, would be forgotten and replaced by fast, economical and automated production methods. Thus, artisans create unique pieces that evidence handcrafted quality, technical skill and high symbolic value.

According to Dresner [8], sustainability has a cultural dimension in addition to environmental, economic and social dimensions. While, broadly speaking, it could be regarded as part of the social dimension, it certainly requires special attention in cases such as the sustainable luxury sector. According to Na and Lambling [18], "For sustainable luxury craft, the ecological dimension refers to the materials and methods of craft-making that minimally impact the environment. The economic dimension relates to all aspects that keep craft businesses and entrepreneurs viable and healthy. The cultural and social dimensions reflect the sustaining elements that keep the values, traditions, and social exchanges of craft alive" [18].

Based on the above, Gardetti points out that "Sustainable luxury would not only be the vehicle for more respect for the environment and social development, but it will also be synonym of culture, art and innovation of different nationalities, maintaining the legacy of local craftsmanship" [11]. Sustainable luxury acts as a bridge between remote communities—with their truly forgotten or unknown wisdom—and the global market to revalue the craft production of these communities.

Moreover, the craft production of these communities—handicrafts—is the vehicle for artisans to express their entire history, wisdom, values and dedication. "Goods embody aesthetic features and production technologies enmeshed in artisans' local traditions" [16: 14]. Artisans' social mobility and financial self-sufficiency are quite diverse. For example, "*vicuñeros*" [vicuña fibre weavers] are aware of the (economic) value of vicuña and its woollen by-products.

In terms of their craft, as years go by they run the risk of becoming lost to the implementation of more efficient methods, or to the discontinuance of certain practices due to their failed introduction into the global market. Therefore, sustainable luxury offers an opportunity to revalue and recover these forgotten customs, giving artisans the possibility of both expanding and continuing with their legacy. In addition to empowering artisans, sustainable luxury supports cultural diversity by showing the world different production methods while enriching the cultural heritage. "The result is that traditional craft serves to meet the social, cultural and increasingly economic needs of emerging economies, but it can also serve to bridge cultures" [15: 392].

The following chapter will focus on *Cooperativa Laguna Blanca*'s vicuña poncho—its handcrafted quality, symbolic character and how sustainable luxury may expand the horizons of social work.

2 The Poncho

2.1 Introduction

The poncho is a simple, functional garment that is part of traditional clothing in different places throughout the world. It was originally worn by American aboriginal peoples. This is a traditional garment of the region, handcrafted in different materials—generally natural fibres—associated with the geographical place and ecosystem where the poncho is created.

The poncho is a classical piece of *criollo* clothing which represents Argentine gauchos. Beyond its use as a piece of clothing, the poncho is attributed to a characteristic role in the country cultural heritage and found in historic photographic documents as represented in pictures, music and literature. As a figure, it encompasses tradition, homeland, culture and skilled craftsmanship. Against this framework, it stands out as an all-encompassing element and, in as far as it preserves its indigenous roots, it survived and consolidated its image during the colonial era.

2.2 The Poncho: History and Cultural Value

While the garment is characterised by shape simplicity, the traditional poncho evidences major technical craft complexity. Today, it is a globally used—and even industrially manufactured—typology. However, it is still very close to its roots. The poncho is considered a "second skin", and it is associated with freedom and simplicity. In Ruth Corcuera's words [5: 163], "In a world where the great urban conglomerates dominate with all their tensions, this garment is associated with the image of freedom and space—an association that perhaps, subconsciously, renders it especially attractive".

This handcrafted textile is made up of weft and warp threads. Two or more threads are interwoven over a frame called loom. The poncho has a rectangular shape, with varying dimensions depending on both the desired garment size and the loom size—reason why it can be woven in one or in several pieces. The finished garment has a central opening for the head and covers (mostly) the (upper) body. While the fibres used to weave it vary based on geographical location, its rectangular shape remains unchanged. These fibres are varied, from animal wool or plants, and different thicknesses. In the north of Argentina, for instance, ponchos are made of linen, silk and cotton. The most popular ponchos are woven from vicuña, llama or ship wool. Men and women have different roles, depending on the area. However, in the Andean world, there are also male weavers, and shearing activities are jointly performed by men and women.

As explained above, the poncho is a garment with deep symbolic character, highly representative of the regional cultural heritage despite its changes over the years. First, it was worn by the indigenous communities of this region. Moreover, in ancient times, woven ponchos were characterised by stripes, drawings, lines, patterns or colours representative and distinctive of a given community. In the same community, families were recognised, and even distinguished, by the motifs on their ponchos. As with fibres, which represent far more than different regions, the poncho material helped identify social status, e.g. the vicuña poncho was only worn by the Inca and his family.

It is a versatile garment that offers free movement while keeping the person warm, or covered, depending on material and weaving, leaving the arms free, e.g. to perform farm and cattle-related activities. For this reason, it was widely used by the gaucho and, when the Europeans arrived and settled in this region, they embraced the poncho as a comfortable, loose garment, characteristic of these lands and its promising future.

2.3   The Vicuña, A Sacred Animal

The colonisation of America chronicles shows that there was a widespread belief that all wild camelids belonged to the Sun. They were considered supernatural, sacred beings which belonged to the deities, and only the son of the Sun and his family were allowed to wear garments made from these animals. The Inca—the son of the Sun—authorised his relatives, or most loyal chiefs, to wear these clothes. The Inca clothes were burnt after being worn by him, so that nobody else could wear them, unless the Inca decided otherwise [3, 4].

The *chaku*[1] was celebrated every 3 years to shear this sacred animal. It was a meeting, a ceremony where vicuñas were captured, and sheared, as well as registered in a census. The fibres obtained were for the Inca, while the meat of some specimens was given to the communities. Since vicuñas have not been tamed, this ceremony still takes place today, but only for shearing purposes, after which vicuñas are released back to nature. In Argentina, this practice takes place in Laguna Blanca, province of Catamarca, and in Santa Catalina, province of Salta, under controlled conditions so that animals suffer the least possible.

The vicuña is an animal of the Camelidae family, native to the Latin American region, especially the Andes. It is the smallest animal in this family, with an estimated weight between 40 and 50 kilos. Just like the guanaco, it is a wild species. The vicuña lives in the highlands, over 3000 m above sea level. At those heights, the soil must offer two essential—though not simultaneous—conditions: on the one hand, it should offer tender grass and, on the other, arid grounds. Vicuñas feed on the grass and they play, turning over the arid soil. The resulting dust is essential to keep their hair loose and fluffy.

Vicuñas are beige or light-brown, with reddish back, and white legs and belly. Vicuña hair is made up of very fine fibres that grow very close together and protect the animal from cold weather, rain and wind. With a diameter of less than 15 μm, its hair is one of the finest fibres in the world. Vicuñas produce between 160 and 180 g of fibre every two years [4, 20].

From pre-Hispanic times, the fibre of this animal has been highly valued, denoting both grandeur and quality. The different shades offered by the hair of a single animal are spun together, thus adding a very special characteristic to the yarn. Due to the fineness of the fibre, several specimens are needed to get a finished product. Both vicuña yarn and poncho production are handmade. In the artisans' hands, this fibre becomes a luxury fibre, since they are capable of turning hair into a fine, delicate yarn. There are few, very knowledgeable artisans who work with this fibre because it is hard to get, very expensive and, difficult to process based on its characteristics. Spinning this fibre requires a high degree of technical expertise in both the processing and characteristics of this material. Only the most gifted spinners, dyers and weavers can do it.

2.3.1 The Vicuña Poncho: Overview

Vicuña ponchos came, particularly, from the provinces of Catamarca, Jujuy and Salta; and even within these regions, there were ponchos woven with a mix of other fibres, such as native and wild silk. Vicuña wool is regarded as one of the most exclusive hair fibres. According to Geijer [12], animal species with the finest hair can be found in the highest mountain regions worldwide. These are cashmere goats, angora goats from Tibet and Pamir Mountains and the shy vicuña from the Andes. Nowadays, vicuña fibre can be mainly found in Peru and, to a lesser extent, in Argentina.

---

[1] *Chaku*: a rounding up, a way to hunt wild animals.

While this fibre is used to make other items, such as scarfs and throws, the most famous garments are the ponchos made in the region of Belén. "At the end of the eighteenth century, the most prestigious ponchos were those confectioned with the precious vicuña wool of Belén" [5]. The distinctive features of these ponchos have always been softness, warmth and lightness, since threads produced from vicuña wool are very fine due to the characteristic animal hair. For example, a vicuña poncho made in Belén weighs less than 225 g and is one of the warmest ponchos [4].

Moreover, this fibre is still positioned among the most exclusive yarns thanks to highly skilled artisans, who chose to continue with their tradition despite the advent and quick, widespread adoption of cotton and other fibres.

3 The Vicuña Poncho of Cooperativa Mesa Local Laguna Blanca

3.1 Laguna Blanca: What Is It? History

The process to weave a vicuña poncho requires history and context. Without such history and context, it is easy to assume that, to make a vicuña poncho, the same effort and processes are used to get the same outstanding result.

For reference purposes, Laguna Blanca is a natural area covering almost 900,000 ha, located in the desert Puna, in Catamarca, 3200 m above sea level (see Fig. 1 and its description). In 1971, Argentina decided to subscribe to the Convention

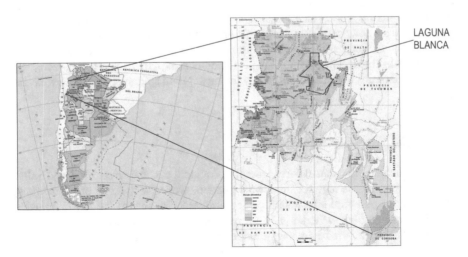

Fig. 1 Location of Laguna Blanca. *Note* On the left, there is a map of the Argentine Republic, without the Antarctic territory and, on the right, the province of Catamarca with the location of Laguna Blanca. *Source Instituto Geográfico Militar* [Military Geographical Institute] (Argentine Republic). Duty-free Maps

on International Trade in Endangered Species of Wild Fauna and Flora (CITES),[2] translated by Law No. 4855. In 1971, the vicuña was declared an endangered species and became part of the CITES.

In 1979, the government of the province of Catamarca declared the area a reserve for vicuña protection. In 1982, the United Nations Educational, Scientific and Cultural Organisation (UNESCO)[3] included the reserve as Biosphere Reserve Category VI of the International Union for the Conservation of Nature,[4] i.e. a managed resource protected area, with pristine natural systems and management of biological diversity and resource sustainability based on community needs.

Laguna Blanca has about 600 inhabitants, including the indigenous communities of Corral Blanco, Aguas Calientes and Laguna Blanca. Besides the home to vicuñas, it is also a shelter for the conservation of birds, such as the greater flamingo, and rich American fauna.

The presence of vicuña and its interaction with human beings can be traced back to ancient times. This is evidenced in the rock art of the area, such as that in Corral Blanco, 12 km away from Laguna Blanca. That is one of the various rock art sites in the Puna. It is a 5-metre-high stone wall that was the medium for artists from the formative stage (2000 BC) to join the forces that interacted with them: the tracks of *surí*–a feline with a spotted coat and camelids tracks. In many cases, Andean rock art evidences a deliberate management of camelid populations in these regions that can be traced back to 4000–5000 years BP—a progressive practice which introduced locals to shepherding.

Archaeologist Delfino [7] ventures to think that, in pre-Hispanic times, Laguna Blanca was a place of interest for caravan groups. And, based on the detailed description of locals' resources and economic practices, spinning and weaving ropes and slings for hunter-gatherers were of utmost importance.

Unlike the llama, as the vicuña cannot be tamed fibre collection requires its capture. Historically, vicuñas were captured through hunt, killing the animal, or through the *chaku*, a method to capture, shear and release vicuñas back to nature.

---

[2]The Convention on International Trade in Endangered Species of Wild Fauna and Flora (CITES) was established in 1973 in response to the growing concerns about the overexploitation of wildlife through international trade, which threatens the survival of many animal and plant species all over the world. CITES has been ratified by 180 countries. The main purpose of the Convention is to ensure that international trade does not threaten the survival of natural species. To this end, CITES regulates international wildlife trade through a system of permits and certificates. Source: Cherny-Scalon [2]. Visit: https://www.cites.org/esp/disc/what.php.

[3]UNESCO is the United Nations Educational, Scientific and Cultural Organisation. UNESCO seeks to build peace through international co-operation in Education, the Sciences and Culture. UNESCO's programmes contribute to the achievement of the Sustainable Development Goals defined in Agenda 2030, adopted by the UN General Assembly in 2015. Source: Official Website https://es.unesco.org/about-us/introducing-unesco.

[4]The International Union for Conservation of Nature (IUCN) is a Membership Union uniquely composed of sovereign states and both government and civil society organisations. The IUCN provides public, private and non-governmental organisations with the knowledge and tools that enable human progress, economic development and nature conservation to take place together. Source: Official Website https://www.iucn.org/es/acerca-de-la-uicn.

Researchers from *Instituto Nacional de Tecnología Agropecuaria* (INTA) [National Agricultural Technology Institute],[5] Gonzalez Cosiorovski and Moity-Maizi [14, p. 65] argue that: "The Quechuan term '*chaku*' means 'hunt' and refers to practices originally attributed to Inca settlements. Ancient texts explain that vicuña hunt or *chaku* took place every three or four years, but there were also other types of hunt (referred to as *chaku* or *qayqus*), reserved to *curacas*, i.e., the community authorities. A large group of men set off to look for vicuña herds in the territory, quickly herding and leading them to a controlled enclosure. Some animals were killed, while others were just sheared and released. Wool was for the Inca and his family. These *chakus* were regulated by political, religious, social, and cultural mechanisms".

The Spanish conquest replaced capture methods, and hunting vicuña for meat, leather and fibres became widespread. Until the mid-twentieth century, Laguna Blanca was a major trading centre of specimens hunted for those purposes. Wild herd population drastically declined to the alarming low number of 2000 specimens countrywide. At that point, the ban on hunting vicuña and the creation of protected areas were the first steps for herd recovery.

The growth of vicuña population was not without challenges. It brought about problems because, in some areas, vicuñas competed for pastures with local cattle. Biosphere resource management emerged as a way "to address the needs" of settlers. This resulted in setting up a group representing different commissions of the same location. "That is why it is called *Mesa Local de Laguna Blanca* [Laguna Blanca Local Board]", explains Ramón Gutierrez,[6] "because it was first made up of the school principal, the police, the municipal delegate and the club president, who gathered to represent and help artisans". The community was aware of Peruvian *chakus* and knew both that their ancestors used those methods and that rock art evidenced interaction with vicuñas. However, they also knew that it was no longer a community practice, since poaching had become widespread.

Since an accountable institution was required for the *chaku*, *Cooperativa Mesa local Laguna Blanca* was created in 2002. The *chaku* is a major production, social and cultural event. Moreover, it is a ritual ceremony that begins with a *corpachada*[7] and propitiatory rites[8] for good shearing. During the *chaku*, all the Co-operative members work hand in hand. According to Gutiérrez,[9] "We put up modules, set a net and work quickly and in an orderly fashion to prevent animal death. In the past five (5) years, not a single animal died during the *chaku*. Five (5) out of one hundred (100)

---

[5]The *Instituto Nacional de Tecnología Agropecuaria* (INTA) [National Agricultural Technology Institute] is a national agency engaged in the sustainable development of the agricultural, agri-food, and agro-industrial sector through research, extension, innovation and transfer of knowledge in Argentina.

[6]Conversations between Ramón Gutiérrez and Roxana Amarilla, July 2019.

[7]As explained by different Andean artisans, *corpachada* is a sacred practice among Andean communities which involve feeding or offering food to Mother Earth.

[8]Propitiatory rites are the practices related to religious observance intended to beg or calm down supernatural beings. Source: Diccionario de Ciencias Sociales y Políticas, Torcuato Di Tella, EMECE, 2001.

[9]Ibid 6.

animals died during the first *chaku*. Now, we conduct *chakus* quickly and in silence, and we have a special net which does not hurt vicuñas when they crash against it; instead, they bounce and we hold them for shearing. In the past, we used tarps, which banged in the wind and scared vicuñas". The first recorded *chaku* took place on 30 November 2003.

After shearing, the sheared fleece is divided: 10% to the owner of the land where the *chaku* took place; 20% to the province, which sells it to other artisans at a subsidised price; and 70% to the Co-operative. Co-operative members split fleece in equal shares. They also buy fleece from the landowner to increase their share. Co-operative members classify and remove bristles[10] from fleece. This is a very important manual process since clean fleece results in subsequent garment softness. Gutiérrez explains[11] that "… we don't sell wool, especially now that we are learning so much from the co-operative movement. Sometimes we buy fleece from the land owner because we use it. Our purpose has never been to sell the raw material but to add value to it; so, we sell a minimum percentage because the Co-operative needs that money to pay annual taxes and for the minimum expenses required for our structure operation. Until three years ago, we used to sell such percentage as shorn, but now we sell it with the bristles removed, creating a source of employment for young people who haven't started weaving yet".

After shearing, classification and bristle removal, yarn production begins. In Laguna Blanca, spinning is an emblematic activity.

Historian Cruz [6] explains that, before revitalising the interaction with vicuñas, the area was visited by intermediaries and weavers to get fine yarns. Besides, Cruz describes an activity called the "vicuña *minga*", i.e. an artisan invited other artisans to her house to help her with vicuña fibre spinning—the most time-consuming stage in the production process. Besides a compensation in money or in kind per spun ounce, the *minga* involved expertise, learning and social interactions between *comadres*,[12] friends and neighbours. The collective labour organisation included meals and usually finished with a symbolic compensation that legitimised the process as a mutual help and co-operation relationship: regional dance and dining. Once the labour at the artisan's house was completed, artisans were available to collaborate with another fellow artisan [6, p. 4].

Nowadays, after yarn spinning, artisans may choose to use—all natural—solid colours or to combine them in a pattern. This is how the raw material transformation stage is complete and they get ready for the weaving process.

Back to Cruz [6], the historian argues that, in the sixteenth and seventeenth centuries, besides weaving textiles for domestic purposes, the weaving population of these regions used textiles as regional trading currency.[13]

---

[10]To remove the bristles means to clean camelid fleece, removing coarse hair so that it becomes softer to touch. Explanation given by different Andean artisans.

[11]Ibid 6.

[12]*Comadre* refers to a neighbour and friend with whom a woman has a closer, trusted relationship than with other women. Source: Diccionario de la Lengua Española, Real Academia Española.

[13]During the colonial era, the regional trading currency [*Monedas de la Tierra*] was used in areas of Paraguay and the River Plate that were far away from the larger centres. Objects used as currency

However, Ramón Gutierrez[14]—president of *Cooperativa Mesa Local Laguna Blanca*—believes that "while *lagucho* textiles have always been top quality textiles, since the recovery of the *chaku* and wild vicuña management, the raw material change forced them to receive new training, for example, in bristle removal and revitalising local textile motifs". To improve textiles, strategies were implemented for knowledge transfer from renowned master artisans such as Sebastiana Salgado, Rosa Salgado, Martina Chaile and Julia Guerra. They also bought wood, as well as loom frames and heddles, to have the technology required for finer finishes and to prevent fibre entanglement. And, then, they recovered designs such as the so-called bird's eye,[15] because, even though artisans have never stopped weaving them, they were only woven by very few people and the Co-operative needed to recover its distinctive designs. At the heart of the group, they give their textiles characteristics a lot of thought. For example, they try to preserve how they use materials. Unlike other *vicuñeros* [vicuña fibre weavers], they weave weft and warp in pure vicuña fibre with no mix of any other fibre. They keep natural animal colours and plan their designs based on them.

This path was taken based on a decision made by all the Co-operative member artisans, who weave at their own pace, with their families, as much as they can. They do not weave upon request.

In 2006, the Co-operative started to attend national and provincial fairs along with other peasant organisations who had also recovered Andean crops. Back then, Andean crops were the most important production. There was almost no sale of vicuña textiles, with only few exhibitions and sales at an information centre made available by the Municipality of Laguna Blanca.

In 2013, seven years later, during the *Fiesta Nacional e Internacional del Poncho* [National and International Poncho Festival][16] the expert jury—made up of textile specialists and historians: Celestina Stramigioli, Manuela Rasjido and Roxana Amarilla—found at the peasant organisations' Andean crop both an excellent pure vicuña fibre textile with a bird's eye pattern and *rapacejo*[17] finishes made by *Doña* Crescencia Casimiro, and an impeccable poncho in graded natural colours woven by young artisan Ana de Lujan Suárez. Both of them are the members of *Cooperativa Mesa Local Laguna Blanca*.

In the same edition of the most important textile festival of Argentina, an engraving offering found in IntiHuasi volcano, 6000 m above sea level, was exhibited for one time only. This offering consisted of a small silver figurine dressed with a cloak and a sash, silver brooches, a basket with camelid fleece and a feathered headdress.

---

varied depending on the place. Source: "La implantación de la moneda en América," Torres, Julio, p. 117.

[14]Ibid 6.

[15]As explained by some artisans, *ojo de perdiz* [bird's eye] refers to a textile design typical of the Andean region.

[16]*Fiesta Nacional e Internacional del Poncho* is an annual meeting and celebration held in the province of Catamarca since 1967 that gathers traditional textile artisans and other regional crafts.

[17]According to different artisans, *rapacejo* or *mallado* refers to lace weaving and textile finishes.

Fig. 2 *Fiesta Nacional del Poncho* 2016, San Fernando del Valle de Catamarca, 18 July 2016. The Argentine Ministry of Culture was present at such national festival. Picture: Romina Santarelli/National Ministry of Culture. *Source* Argentine Ministry of Culture. Duty-free Images

Around the cloak, the figure has an Andean cord with the "bird's eye" design. A sign of ancient times highlighted the presence of an ancient Andean motif recovered by Crescencia Casimiro's top-quality textile. As to Anita Suarez's poncho, besides the excellent weaving technique, the jury especially recognised the quality finishes. Celestina Stramigioli said: "Outstanding, small, perfect stitches, a tiny embroidery reminiscent of old times".[18]

The awards granted to both Co-operative artisans had an impact on the group. "This was more exciting, because it helped us add more value to our textiles", explains Ramón Gutiérrez.[19] Besides, they were able to participate in the areas intended for traditional artisans, and they received the best poncho award thanks to the excellent pieces woven by artisan Zulema Gutiérrez, in 2015; Ana de Luján Suárez, in 2016; and Zulema Gutiérrez, in 2018. In the last edition held in 2018, one of the greatest Laguna Blanca master artisans, *Doña* Julia Guerra, was awarded the First Prize to the Best Sheep Wool Textile Handicraft.

In connection with this experience, Cruz [6, p. 5] thinks that: "In the past editions of the *Fiesta Nacional del Poncho* young women from the Puna who demanded the end of 'how much do you want to pay for my textiles?' with their art as '*vicuñeras*' (Figs. 2, 3 and 4). The jury rewarded and recognised both their works and craft growth,

---

[18]Celestina Stramigioli's speech at *Fiesta Nacional e Internacional del Poncho* the award ceremony in 2013.
[19]Ibid 6.

Fig. 3 *Fiesta Nacional del Poncho* 2016, San Fernando del Valle de Catamarca, 19 July 2016. Within the framework of the *Fiesta Nacional del Poncho* 2016, a training in co-operatives and association movement for artisans was delivered by Graciela Correa, *Instituto Nacional de Asociativismo y Economía Social* [National Association Movement and Social Economy Institute]. Picture: Romina Santarelli/National Ministry of Culture. *Source* Argentine Ministry of Culture. Duty-free Images

as well as 'the artisan's quality' as an expression of the individual and collective know-how, which is consolidated, suitable, verifiable and recognised by other non-artisans. Julia Guerra, Crescencia Casimiro, Ana Suárez and Zulema Gutiérrez's vicuña ponchos and throws evidenced that their designation as mere 'wool' sellers imposed for decades by intermediaries and even by artisans in the valleys had come to an end".

This recognition was consolidated when a post of the Loom Route—along route 40—was set up in Laguna Blanca. Instead of resting on its laurels, the Co-operative recently achieved another extremely important milestone: the internationalisation of its trading area.

3.2 The Experience in Cusco

In 2017, the Co-operative was invited by Doña Nilda Callañaupa to participate in the weavers' Tinkuy 2019, in Cusco (Peru). This gathering is organised by the Centro de Textiles Tradicionales *del Cusco* [Cusco Traditional Textiles Centre], a key benchmark institution that managed to bring together traditional textiles, quality and traditional artisan organisations. Young spinner María Gutiérrez and Ana del

Fig. 4 *Fiesta Nacional del Poncho* 2016, San Fernando del Valle de Catamarca, 18 July 2016. The Argentine Ministry of Culture was present at such national festival. Picture: Romina Santarelli/National Ministry of Culture. *Source* Argentine Ministry of Culture. Duty-free Images

Lujan Suárez, the artisan who received two awards at the *Fiesta Nacional del Poncho*, travelled to Cusco with the support of the National Ministry of Culture. There, not only did they show their textiles, but they also described the experience of the Co-operative's organisation and growth.

3.3 From the Heights of the Puna to New Mexico Collectors' Market

In 2018, through the *Mercado de Artesanías Tradicionales de la Argentina* (MATRA) [Market of Argentine Traditional Handicrafts],[20] they managed to become part of the curatorship of the prestigious International Folk Art Market (IFAM),[21] the traditional handicrafts fair held once a year in Santa Fe city, New Mexico, which gathers luxury collectors, curators and consumers. This was a positive experience and the

[20]Mercado Nacional de Artesanías de Argentina (MATRA) [Market of Argentine Traditional Handicrafts] is the National Secretariat of Culture's handicrafts programme which, among other missions, works to connect artisans to the international handicrafts fair market.

[21]International Folk Art Market (IFAM) is an annual traditional artisans' fair which also has an empowerment and incentive programme for artists worldwide.

Co-operative sold its fine textiles. This year 2019, the Co-operative will participate once again in the IFAM, but this time in a highly visible pavilion.

Moreover, and as a part of the pop-up store of a small group of six (6) traditional artisans in the exclusive Urban Zen, IFAM also selected designer Donna Karan's Manhattan store.

In these past international participations, garments travelled with special stamps jointly coordinated by the provincial and the national governments to certify fibre origin, type and sourcing method.

Cooperativa Mesa Local Laguna Blanca has vast experience, leaving its sustainability footprint which involves an all-encompassing concept consisting of resource management, revitalisation of good environment-friendly and cultural community practices, textile diversity and weaving quality, co-operative movement and the collective as a tool for development, community identity and certainty of a better future.

Therefore, the production process of *lagucho*[22] vicuña ponchos can only be understood within the community development history of this indigenous people.

3.4   Some Institutions and Organisations that Supported the Co-operative and Consolidated Its Entry into the International Market

Provincial, national and international programmes support the Co-operative and its entry into the international market.

Firstly, at a regional level, we should mention Catamarca Provincial Biodiversity Directorate. Among other functions, it is responsible for ensuring biodiversity conservation, promoting and monitoring the enforcement of regulations on management, conservation, preservation and sustainable use of different natural resources—fauna and flora—of the province; and it is the enforcement authority of laws on management, preservation and sustainable use of the natural ecosystems and environments of the province.[23]

There is also the Provincial Handicraft Directorate, which is responsible for monitoring and promoting the management and implementation of public policies intended for artisans who live in the province. It promotes two large handicrafts fairs: the *Fiesta Nacional del Poncho* and the *Feria Manos Catamarqueñas* [Catamarca Hands Fair].[24]

---

[22] As explained by the locals, *lagucho* refers to the people and objects from Laguna Blanca.

[23] Source: Official Website, *Ambiente y Desarrollo Sustentable* [Sustainable Environment and Development]. http://www.ambiente.catamarca.gov.ar. Accessed: July, 28.

[24] Source: Official Website, http://www.artesanias.catamarca.gob.ar Accessed: July, 28.

Moreover, there is the *Museo Integral de Laguna Blanca* [Laguna Blanca Integral Museum] consisting of an Interpretation Centre, an interpretive trail with an onsite museum, a botanical park, rock art sites and the entire biosphere reserve.[25]

Secondly, at a national level, the *Instituto Nacional de Tecnología Agropecuaria* (INTA) [National Agricultural Technology Institute] is a state-run decentralised agency with both operating and financial autonomy that depends on the National Agro-industry Secretariat. It regards innovation as the driver of national development, focusing on capacities for the agro-industrial sector and creating networks to promote interagency co-operation. In turn, it generates knowledge and technologies made available to different sectors of society through its outreach, information and communication systems.[26]

Moreover, the *Mercado Nacional de Artesanías* (MATRA), a handicrafts programme of the National Secretariat of Culture, was created in 1985. It focuses on handcrafted quality, fair trade and price, visibility of Argentine artisans' work and sector strengthening in the creative economy. In turn, it is the committee of the World Crafts Council's Recognition of Excellence in Handicrafts for the Southern Cone and focal point of the Ibero-American Handicrafts Programme.[27]

Thirdly, on a global basis, the Co-operative is supported by the *Centro de Textiles Tradicionales de Cusco* (CTTC) [Cusco Traditional Textiles Centre], a non-profit organisation that promotes weavers' empowerment through revitalisation and sustainable practice of ancient textiles from the Cusco region. Its mission is to help weavers preserve both weavers' identity and textile traditions, while improving their quality of life through education and promotion of their textile art.[28]

The Co-operative handicrafts are also present at the International Folk Art Market (IFAM), a traditional handicraft fair and programme operating since 2004 in Santa Fe city, New Mexico, USA. It is the world's largest market of its kind, supported by non-profit empowering international folk artists. The mission of the IFAM is to create economic opportunities for and with folk artists worldwide who celebrate and preserve folk art traditions.[29]

Further to the above, the Co-operative was also supported by other initiatives.[30]

---

[25]Source: Cultural Promotion Website, https://ilam.org/index.php/es/museo?id=6242. Accessed: July, 27.

[26]Source: Official Website, https://inta.gob.ar. Accessed: July, 27.

[27]Source: Programme's Facebook page "MATRA-*Mercado Nacional de Artesanías Tradicionales de la Argentina.*" Accessed: 26 July, 2019.

[28]Source: Official Website, http://www.textilescusco.org. Accessed: 29 July, 2019.

[29]Source: Official Website, https://folkartmarket.org. Accessed: 27 July, 2019.

[30]Other initiatives supporting the Co-operative:

Universidad Nacional de Catamarca (UNCA) [National University of Catamarca], which is committed to the culture and identity of Catamarca—and its promotion—through education, research and outreach. The institution revalues the tangible social and cultural heritage, further networking with the public and private social and production sectors. Source: Official Website, http://www.unca.edu.ar. Accessed: 29 July, 2019.

Along this line, the purpose of the *Instituto Interdisciplinario Puneño—Universidad Nacional de Catamarca* (InIP-UNCa) [Interdisciplinary Institute of the Puna - National University of Catamarca]

4   Sustainable Luxury and Indigenous Production

Indigenous practices have been replaced by automated practices dictated by an economic model that pursues development based on capital increase at any cost, while marginalising and dividing society. In this case, Laguna Blanca community evidences how the integration of ancient culture, cultural heritage and sustainable development can coexist and leverage each other as well.

According to the UN,[31] "Indigenous peoples are inheritors and practitioners of unique cultures and ways of relating to people and the environment. They have retained social, cultural, economic and political characteristics that are distinct from those of the dominant societies in which they live. Despite their cultural differences, indigenous peoples from around the world share common problems related to the protection of their rights as distinct peoples". Precisely, the commitment of sustainable luxury to these communities is essential to protect their rights, but also to expand their frontiers. It must ensure respect, protection and appreciation of the regional cultural wealth.

Nowadays, these communities maintain the social indigenous experience as they try to combine ancient knowledge with globalisation, the advancement of technology and the global community. No doubt they intend to remain faithful to their ancestors by teaching these practices to young people—the importance of sustainable luxury here lies, as it is the way to keep all the above alive.

4.1   *Characteristics of the Social and Production Community Experience*

Some artisans' organisations, such as co-operatives, improve their possibilities when they enter new markets. Since they are self-managed organisations, they share not only production techniques, but also their ideals, values and history. One of the main characteristics of co-operatives is that they encourage participation and equality,

---

is to develop, foster and promote scientific research of the Puna region; support the training of researchers; and direct the interrelation between research, education, university outreach and both regional and local development. Source: Official Website, lagunablanca.unca.edu.ar. Accessed: 27 July, 2019.

There is also, Becar, a National Secretariat of Culture international grant programme developed on the basis of Co-operation for Artistic Education, Research and Creation. It supports international mobility for artists and cultural professionals to carry out artistic projects abroad. Source: Official Website, https://becar.cultura.gob.ar. Accessed: 27 July, 2019.

Finally, until December 2011, the Co-operative was supported by the *Proyecto de Desarrollo Rural del Noroeste Argentino* (**PRODERNOA**) [Argentine North Western Rural Development Project], a project to invest in rural production and service activities that leveraged the resources available to male and female small farmers from vulnerable groups. Source: Official Website, https://www.ucar.gob.ar/index.php/prodernoa. Accessed: 28 July, 2019.

[31]For more information, please visit: https://www.un.org/en/events/indigenousday/. Accessed: 8 May, 2019.

while striving for a common, concurred goal, as explained by Grimes and Milgram [13]: "Most artisans groups desire to preserve local values through their production. They are not working solely to market products but also sustain deeply held beliefs about their social relations and their relationships to the environment" ([13]).

Organised artisans are able to present their "savoir fair" and (ancient) culture in the global market to become known. And sustainable luxury may be—and truly is—a factor that would help artisans achieve deep market penetration.

According to Causey [1], a co-operative can only work and grow if it manages to be represented and promoted in the global market [1].

Indigenous people who live in these regions regarded the vicuña as a sacred animal and even though today there is no such belief, animals are truly respected in their wild natural environment and the *chaku* is still celebrated. While the members of Laguna Blanca community do not consider themselves aborigines, they maintain their customs and, above all, respect their ancestors. Thanks to the creation of the co-operative, they gather, share their wisdom and manage to broaden their horizons and address their problems while strengthening their sense of belonging and values.

In this case, through the vicuña poncho, *Cooperativa Laguna Blanca*'s artisans share their vision of the world, their rich history and, above all, they teach about cultural value and tradition with a competitive object in the sustainable luxury sector. And, precisely, consumers in this sector wish to go beyond the consumption of standardised products. According to Stephen, sustainable luxury consumers look for authenticity and pursue quality. "The international market for crafts is built on an elite consumer ideology that contrasts manufactured, mass-produced, modern objects with handmade, authentic, local crafts" [19].

Regarded as one of the world's most exclusive fibres, vicuña wool is certainly a luxury fibre. As it is a wild animal, its shearing—based on the ancient aboriginal method that respects both the species and its origins—is considered sustainable. Not only does this practice continue with the indigenous tradition, but it also enriches the sustainable luxury universe, as it presents a history, a tradition, preserving culture with a practice that comes from times immemorial.

5 Some Considerations and Conclusions

Sustainable luxury is recognised as a meeting point for traditions, cultures, values and needs. It rescues and expands the cultural heritage of communities, enhancing their history to share it with the world. Craftsmanship is at the heart of sustainable luxury because artisans are the ones who add value to objects.

While today the poncho has become a popular garment found in many international collections, and—to a certain extent—losing its symbology, Laguna Blanca artisans are the ones who add values, traditions and culture to this item. Handcrafted ponchos woven in the Co-operative with the natural fibres sheared by its members still retain their character and hallmark. According to Corcuera: "In the hands of this creole weavers, the poncho still retains the resonances of archaic rites and inspires in our imagination a memory of Eden" [5].

Artisans, their life stories and their creations manage to turn these objects into sustainable luxury pieces; and it is precisely handcrafted practices that distinguish authentic from mass luxury since customers perceive and value the quality of a handmade object [9].

The poncho is part of the regional cultural heritage, with vicuña fibre as its most exquisite example. No doubt the symbology, technique and materials underlying this creation turn Laguna Blanca and its community into sustainable luxury artisans. As mentioned above, a few years ago, the vicuña was an endangered species. However, species preservation is now ensured thanks to artisans like those in Laguna Blanca, who managed to become known and teach their indigenous culture in the sustainable luxury universe. Hence, they can continue with their ancient tradition, enriching the global sustainable luxury market and teaching the value of their culture.

References

1. Causey A (2000) The hard sell, Anthropologists as brokers of crafts in the global marketplace. In: Grimes KM, Milgram BL (eds) Artisans and cooperatives. The University of Arizona Press, Tucson, pp 159–174
2. Cherny-Scanlon X (2014) The wild side of luxury: can nature continue to sustain the luxury industry? In: Gardetti MA, Torres AL (eds) Sustainable luxury: managing social and environmental performance in iconic brands. Greenleaf Publishing, Sheffield
3. Corcuera R (1987) Gasas prehispánicas. Fundación para la Educación, la Ciencia y Cultura, Buenos Aires
4. Corcuera R (2017) Ponchos de América. De los Andes a las pampas. Fundación CEPPA Ediciones, Buenos Aires
5. Corcuera R (2005) Ponchos of the river plate: nostalgia for Eden. In: Roots R (ed) The Latin American fashion reader. Bergpublishers, Oxford, pp 163–174
6. Cruz R (2005) "Ponchos Catamarqueños." Buenos Aires, Ministerio de Cultura de la Nación, inedito
7. Delfino DD (1995) Arte Rupestre e Interacción Interregional en la Puna. Buenos Aires, Cuadernos del Instituto Nacional de Antropología y Pensamiento Latinoamericano 16:367–400
8. Dresner S (2002) The Principles of Sustainability. Earthscan, London
9. Gardetti MA (2018) Lujo Sostenible: Creación, desarrollo y valores de una marca. LID, Buenos Aires
10. Gardetti MA, Rahman S (2016) Sustainable Luxury Fashion: A Vehicle for Salvaging and Revaluing Indigenous Culture. In: Gardetti MA, Muthu SS (eds) Ethnic Fashion. Springer, Singapore, pp 1–19
11. Gardetti in Conference dictated at the seminar sustainable luxury and design within the framework of the MBA of IE. Instituto de Empresa, Madrid, 2011
12. Geijer A (1982) A history of Textile Art. Pasold Research Fund-Sotheby Parke Barnet, London
13. Grimes KM, Milgram L (2000) Introduction, Facing the challenges of artisan production in the global market. In: Grimes KM, Milgram L (eds) Artisans and cooperatives. The university of Arizona Press, Tucson, pp 3–10
14. Gonzalez Cosiorovski J, Moity-Maizib P (2019) La Fábrica Patrimonial Del Chaku en la Reserva de la Biósfera de Laguna Blanca, CATAMARCA, Argentina, Tétralogiques, No 24, pp. 59–74

15. Hawley JM, Frater J (2017) Craft's path to the luxury market: sustaining cultures and comunities along the way. In: Gardetti MA (ed) Sustainable Management of luxury. Springer Nature, Singapore, pp 387–410
16. Littrell MA, Dickson MA (2010) Artisans and fair trade—crafting development. Kumarian Press USA
17. Pop M (2016) Sustainability and cultural identity of the fashion product. In: Muthu SS (ed) Gardetti MA. Springer Nature, Singapore, pp 83–104
18. Na, Lambing (2012) Sustainable Luxury: Sustainable crafts in a redefined concept of luxury from contextual approach to case study. https://www.plymouthart.ac.uk/documents/Na_Yuri_Lamblin_M.pdf
19. Stephen L (1996) Export markets and their effects on indigenous craft production: the case of the weavers of Teotitlan del Valle, Mexico. In: Schevill MB, Berlo JC, Dwyer EB (eds) The textile traditions of Mesoamerica and the Andes: the anthology. University of Texas Press, Austin, pp 381–402
20. Vilá B (2001) Camellos sin joroba. Colihue, Buenos Aires

Unwritten: The Implicit Luxury

Cindy Lilen Moreno Biec

Abstract Can materials make us communicate better and deeper, without words? Ancestral textiles held such a power, but many of these have vanished despite the attempts of some tribes to keep those secrets alive. Our task is to help bring back that knowledge and a haptic way of thinking in order to fight the culture of prohibiting touch which has evolved in many places. Touch is the sense that holds the most memories, opening new worlds of communication connecting hand, eyes and brain. This paper will analyse how artisans and craft can help to make this happen via three ancestral textile languages—the Andean Quipu system, Bogolanfini mud cloth from Mali, and Mongolian felt. Taking these unwritten languages as metaphors for human development and interaction, and understanding that humans have found different ways to express themselves when the possibility to translate everything into words does not exist, we see that workmanship is a central concept for all of them. For those societies, these languages were a means to create their own worlds, keep memories, develop their identities, and make their thoughts and cultures last. These textiles, being holistic, can go beyond boundaries and achieve better interactions because they do not rely on intellectual knowledge, instead engaging feelings and senses, producing internal wellbeing which can be later passed on to the external world. Recovering these languages and the way of thinking they inspire helps us to reach a state where we create our own identity beyond the culture around us, and closer to life experiences, places, and knowledge. These environments would be designed with natural materials and encourage users to be part of that space using their hands to adapt everything to their needs, to be part of the journey of their senses. In a world where we often communicate in a very impersonal way, using technology, and sometimes feeling isolated even when surrounded by people and things, being consciously present in the moment is now the new luxury. Can natural materials and craft overcome this and offer humans a better way of interacting with others and the world them? They probably can, but this is not a luxury that can be bought; it requires people to put brain, hands, and soul together to make it work.

C. L. Moreno Biec (✉)
Patagonia, Argentina
e-mail: info@cindylilen.com

© Springer Nature Singapore Pte Ltd. 2020   45
M. Á. Gardetti and I. Coste-Manière (eds.), *Sustainable Luxury and Craftsmanship*,
Environmental Footprints and Eco-design of Products and Processes,
https://doi.org/10.1007/978-981-15-3769-1_3

1 Introduction

Brené Brown, in her TED talk discussing vulnerability, said that over 10 years of researching human behaviour, she has found that everybody needs connections, so much so that feeling connected is a basic human need; connections of belonging, human connections, and connections with the environment. Good connections are directly linked to communication, but is it possible to improve them, and find a way to make them deeper and more meaningful? When we talk about communication we immediately think about spoken and written languages, but these languages have limits. Not everybody speaks the same tongue, and even if they did, there are aspects of human experience that spoken and written languages cannot reach. So what about unwritten languages? Something that could engage not just our intellect but also our senses and even more that could make us feel present in the world and interact with one another beyond boundaries? Looking for different ways of communicating that can go deeper that can let people interact in a more sensitive way, and we cannot ignore the importance that textiles have had throughout history in achieving this. This paper is going to analyse how ancestral communities developed unwritten languages through textiles and how understanding, engaging with, and preserving the dynamic they used can help us connect with others in the deepest way. This type of connection can be considered a new form of luxury, one that is no longer about buying things but about understanding and reacting to the needs of both people and the environment in a holistic way.

To understand how people in the past used textiles to create their own worlds, keep memories alive, develop identities, and make thoughts and cultures last, we will break down the characteristics of three unwritten ancestral textile languages: The Andean Khipu system, Bogolanfini Mudcloth from Mali, and Mongolian felt. These three different cultures can show us how they had to reach a more complex language when words were deemed insufficient to communicate with others and understand the world and its complexity.

Recovering these languages and the way of thinking they inspire helps us to reach a state where we create our own identity beyond the culture around us, and closer to life experiences, places, and knowledge. In a world, where we often communicate in a very impersonal way, using technology, and sometimes feel isolated even when surrounded by people and things, being consciously present in the moment is now so infrequent that it can be considered one characteristic of the new luxury. Heidegger [6] said that what makes us human is "being in the world", which we can do by interacting not just with one another, but with the environment and the objects around us. Objects and our connection to them make us present, much as Steiner has described: "To be human is to be fixed, embedded and immersed in the physical, literal, tangible day to day world" [10]. Knowing the response of our bodies, senses and minds to different stimuli lets us understand how we connect with others, the world, and the vast and complex map of human interactions.

The new luxury is also based on what is happening in our surroundings. The way to experience it now is through our bodies, using the senses, going to feel, to touch;

but how can we get there if in many places a "do not touch" culture has evolved, prohibiting touch and thus limiting our body's perceptions? Touch is the sense that holds the most memories, and when we make it react we open a whole new world of communication. Lesley Millar says in *Surface as Practice* [8] that we go to touch to make others "feel reassured, cared for and closer to us. Expression on a physical level can bring two identities together and generate extremely powerful responses". Making with our hands, uniting touch with materials, generates levels of well-being and satisfaction in much the same way.

When discussing the connections humans need, we cannot avoid our relationship with nature, because part of the feeling of well-being is related to our engagement with nature. Even if we consider humans to be the most developed species, we are part of a greater whole; "we are all made of each other" [7]. The link between exposure to nature and well-being is so well established that growing trends such as Biophilic Design have started to place more emphasis on including elements of nature in architecture and interior design. The possibility that natural materials can help this dynamic of improving happiness and well-being in a silent but effective way, feeds into the increased importance being placed on protecting the environment. Acting to safeguard the future of the world is not something that we can continue to delay, which is why a culture that cares about sustainability is on the rise. This sustainability is starting in simple and individual acts but is rapidly being passed on to broader production systems and larger corporations. Being engaged with the preservation of nature at all levels is part of a new mindset, it is a change, whereby we are looking to improve ourselves not just as individuals, but to reach wider parts of society and the environment.

This is just one of a series of major changes that we are currently going through. As Daniel H. Pink mentions in his book *A Whole New Mind*, we are leaving the era of intellectual and logical thinking behind, and entering an age built on creativity and empathy. This new era is about the "capacity to detect patterns and opportunities, to create artistic and emotional beauty, to craft a satisfying narrative, and to combine seemingly unrelated ideas into something new". Through this, he argues, we will improve our ability to "understand the subtleties of human interaction, to find joy in one's self and to elicit it in others, and to stretch beyond the quotidian in pursuit of purpose and meaning".

This change in mindset means it will be crucial to engage with what is happening around us and take the right actions to incorporate the wisdom that ancestral communities have preserved for us through their techniques. The era of intellectual and logical thinking that Pink mentions helped create a culture that is based on instant technology and consumerism, but at the cost of detaching us from slower ways of engaging with the world, and by extension nature, through our senses. Preserving or resurrecting techniques of craftsmanship can help to find that lost connection with our senses, and offer us a new perspective about how we can use it as an advantage in today's society.

Thus, natural materials, sustainable production, and craftsmanship help us to find purpose and meaning and offer a pathway to reach this new luxury. However, this luxury cannot simply be bought; it requires people to absorb this new mindset to develop the ability to be present in the world, and garner the well-being that comes from it. If this is something that we cannot buy, how can we know what we need? Firstly, we need to understand how we can engage with the world.

This is the moment to stop putting everything into words, stop touching only screens, and start experiencing the world in a haptic way, because today we can access luxury not just with money but with our own senses and mind. The question is, is it possible to learn how to do so? This is where it becomes necessary to look into the role of ancestral textiles as communicators and the holistic world of connections that they hold.

2   Khipu

The Khipu system, a knotting language created by the Incas in Latin America, was a form of both recording data and storytelling, preserving their culture through these knots. This system is so complex that it has never been fully translated, but Khipu expresses its own form of beauty once one becomes familiar with it, and that beauty is in its tactility. The scope of this language is vast, having been used to record universal matters like cities, constellations, time, planets, and so on; all the knowledge of the universe embodied in knots. Khipu is a visual language that speaks of its own process, to name things that cannot be named.

Nowadays, these Khipus take a form of art to tell stories. Time wears out these fragile spaces and does not succeed in keeping alive the memory of everything that has happened in these places. There might exist a preconception in the mind of the user according to which it is impossible to know precisely the stories told by Khipus because of the multiple possible variables. This way of telling stories in fact presents many physical similarities between different pieces, and one could even think that the knots are all "identical", but the difference lies in the vision that each one gives to it. Also, another important fact is that in the Inca Empire only the Quipucamayocs, a class of trained specialists in Khipu, had the knowledge of how to read them, which means that even for many people of its own society this language was a mystery. This is sometimes the main obstacle for unwritten languages: because the meanings they hold are not implicit in any society, they always need someone willing to learn about them and others willing to be taught. It is a knowledge that needs engagement from both parties to be passed on. For the Incas, it was only available for a certain part of society.

Two of the most relevant Khipu artists are Jorge Eielson and Cecilia Vicuña, both visual artists and poets, finding in poetry the closest language to emotions. Eielson made Khipu the main resource for his exhibition *On the Other Side of Languages* [12] where he takes this system to another level showing the limits of conventional languages, putting in Khipus made of fabric those unspeakable matters. He uses these knots as a mechanism of emotional expression. Tarragon and Restany analyse Eielson's work as based on different codes that merge into a single language: "The integration of this continuous event that is the universe; the extreme fluidity of the realm of language translates into a vital nomadism, multiform from the standpoint of expression, planetary from of standpoint of existence" [16]. This knot is for Eielson his form of expression and being in the world, reflecting his own experience in this unique way to present his vision of his realm and life as different to many others.

In contrast, Vicuña uses Khipu to show the emptiness experienced when a whole language and culture is lost, together with our capacity to understand different, more tactile, languages. For her, the line (cord) is a trail of communication. She is constantly weaving words and Khipu together to show the loss of meanings when translating into another language. She links Khipu with nature to show the deep connection that this culture had with natural matters like constellations, stones, mountains or rivers; thus, explaining that our limitations in understanding these languages come perhaps from our lack of connection with the natural world. Vicuña believes that textiles can be read as active texts that play out the on-going intercultural dialogue of self-determination.

The research into Khipu has been taken further by Quilter and Urton [9] explaining not just the use and meaning of these knots, but also the different interpretations through the years. They figured out that black strings signified time and three-plied fibre of a mixture of yellow, blue, and white symbolised gods. With this information, it becomes clear that the purpose of this language was not to record things or have a symbol for every word but to capture things and their whole meaning. Khipu was a language not based on an alphabet but on narratives, thus having a strong link with memory and identity.

Sheila Hicks, another artist inspired by Andean cultures, mentioned that these people mastered the three dimensions and the material creation, and highlighted that they understood the importance of interacting with fibres and structuring thoughts with threads and lines. "What does touching the material add? It connects hand, eye and brain, thus building a triangulation that comes from passion, heart and intellect inseparably cemented to your times and to your own emotional experiences" [14].

3 Bogolanfini

Bogolanfini, from Mali, Africa, is a compound word comprised of three words of the Bambara language: bogo means "earth", lan means "with", and fini means "cloth"; Mudcloth. The making of this particular textile is time-consuming as it is entirely handmade. It is a very long process that can take up to four weeks depending on the weather. This language is based on local knowledge about natural materials and symbols, and the way in which varying natural factors could lead to an uncertain result. Men weaved the cloth and women printed the designs. Women printed them with symbols that only they could understand, thus providing a form of internal gender communication that gave them power and strength over men.

Sarah C. Brett Smith, in her text *Mudcloth: A women's art in Mali* [3], mentions that in the past none of the symbols used in this cloth were put into words; textiles and gestures were sophisticated languages that allowed men and women to come together and stay apart without compromising each other's sources of power. This language was a reaction to a culture where direct speech was traditionally frowned upon; they developed symbols to conceal knowledge.

Many of the symbols talk about simple everyday things that are important for this society. Also, understanding nature—the dying properties of mud, the effects of inclement weather on the various stages of the process, the time needed for processes that are beyond human control—is vital in this language, thus becoming a strong memory holder and identity developer. For many, these pieces of cloth have also the ability to heal, and by mixing certain symbols and colours they developed a way of deep connection between the user, the maker, their beliefs, and nature, thus making a formula with magical and therapeutic properties.

When a language has so many layers of meaning it can enhance people's well-being in a very discreet, everyday manner that is more effective than a more ostentatious equivalent. It is also interesting to observe that in the making of the dye the primary resource is soil, the Latin word for which is "humus" which is also the root for "human" and "humility". The connection with the soil makes us both humble and human.

Satish Kumar, an Indian activist and ecologist, concluded in his book *Soil, Soul and Society, a New Trinity for our Time* [7] that soil (mud) represents life on earth. For a digital society like ours, it is very important to re-connect with the soil and to touch it. It is a way to be part of a healthy web of life. If we recover this interaction with nature, slowing down and using our senses, this will improve self-realisation, and taking care of ourselves is how we can take care of the world.

4 Felt

The third non-verbal language, Mongolian felt, is one of the oldest forms of textiles, dating back 8500 years, and its particularity resides in its process. In places like Mongolia, the making of felt was a communal activity, whereby people congregated in a sacred ceremony where they not only exchanged their skills but also shared the moment of creation, building a strong tradition which helped to keep their culture alive. Felt is one of the simplest textiles, for which the fibre does not need to be spun or woven; it is made from fibres in their natural state only adding heat, pressure, and agitation. It is a textile that can be made in the most remote areas and with few resources, which is why felt has always been important for nomadic peoples. The advantages of felt in terms of protection for basic needs, shelter, warmth, and insulation made life easier for many people.

Mongolians created their own felt houses called yurts. Having made felt the chosen material to surround their lives with, these people believed that felt was what made life possible. This community put a lot of effort into making the yurts because for them, most of whom were nomads, it was the only place to feel at home and protected from the changing surroundings, and for this reason, they have countless hidden meanings. In the middle, a crown made of wood formed the top structure from which the yurt was hung, symbolising the sun that comes through it and the centre of the world. They also have a square piece of felt which works as the main door, signifying home and safety.

The importance of felt resides not just in its simplicity and the high level of nature embodied in it, but also in its anatomical associations. Felts has a natural touch, it can be re-shaped, and has been used for shoes and hats. It is a material that has always been very close to the human body, not just to protect it but to exchange deep emotional qualities; humans added friction and heat to the fibre in order to create the felt in the first place, and, in return, the felt protected human bodies and offered them a sense of belonging and comfort.

One artist who has worked with felt is Joseph Beuys who tired of art forms that were disconnected from everyday life and decided to make art using so-called poor materials from everyday life. Using a material that can be found in almost any household was a way for Beuys to talk to everyone, and thus to develop his own universal language. For Beuys, art needed to engage with "elementary forgotten knowledge" [13]. He wanted to show this basic but irreplaceable elemental material which has protected humans throughout history and so bring back those forgotten connections to simple and common things. He also believed that raw materials provide matter for thought [13]. According to Borer, when asked "who is qualified to create?" Beuys would answer: "those who know the language of the world, that is to say, you and I..." (1996). This is the hidden treasure, improving our knowledge of "sensual languages", those with no need for words, the most universal. When the ineffable is attained, we become challenged, but we need to learn to explore that challenge via emotions instead of our normal, but in this case futile, method of exploring only in words.

5 Why Non-verbal Textile Languages?

These languages were able to engage hands, body, and mind together. Modern observers require a new way of expression and of imprinting their own stamp of their perception of the world. According to the Aristotelian model, "memory is of the past, perception is of the present and conception is of the future" [1, pp.47–48]. As Quilter and Urton said in their book *Narrative Threads*, when analysing the Khipu system and considering the knowledge of ancestral cultures in the past, we can analyse and give sense to the present, and in so doing be better able to project towards the future. Bringing this knowledge back and adapting it to present needs is how we can smooth the way to the future. Being aware of the knowledge that our ancestors acquired, and using it in a more developed way, adapted to today's society, is a form of evolution. It is always good to be innovative, but if we do so considering what has gone before, the achievement of improvements is going to be greater, and more useful for us. Offering new spaces of reflection for observers in the search for active roles are a way of facing the future.

By way of example, these three cultures put their knowledge and heritage into the making, feeling the language through senses. As Maria G. Cattel and Jacob J. Climo mention in their book *Social Memory and History* [4], "Memory is the foundation of self and society". Textiles, like humans, have memories; they tell us about what they went through. We can see examples of this in the dying during the Mudcloth process, which at first is invisible and then starts to slowly reveal the design that holds the meaning of the trace left by the mud, or in the felt that can be re-shaped according to need.

Felt also engages memory as it is a tactile material that impacts on our body, consciousness, and environment. During felting, everything that was in the air at that moment (like dust or hair) gets absorbed into the final piece, making every felted cloth unique. When a fabric impacts memory it engages the social life of things, thus transforming material items into objects that have an active life as people engage with them in multiple meaningful ways [11, p. 163].

When we experience surfaces through our senses we perform a "memory back-up" because bodily memories, which leave marks on our skin, can guide us when conscious memory fails. Those memories are kept through the story and culture that the skin already carries. That means that the same haptic experience will not have the same impact on two different individuals.

Memory is not just an intellectual action, it is also shaped by imagination. Through time, we are transmuting our memories following our imagination and perceptions. That is why engaging our whole body in what we do is a more effective way to preserve meanings and to imprint a long-lasting perception. The richer the engagement, the more holistic and faithful the memory.

However, textile languages do not need to be fixed in order to generate such powerful memories; they can be movable, "unknotted", re-shaped or re-signified. It is specifically their tactility which produces the deep engagement between the "author" of the piece and the end user, and using this sense to "read" creates a personal engagement with the material. The felting process also brings an understanding of versatility and malleability. Using the fibres in their natural state generates a good channel of interaction with nature, and there resides the simplicity of this language. It is an almost magical process to see the fibres coming together to create a new, strong, unwoven surface that can be re-shaped and adapted.

With the right stimuli, humans reveal their identities and allow themselves to be re-signified as well. The materials that act as the conduit for these languages engage human senses, producing internal well-being, which can later be passed on to the external world. As Neil Harbisson has said, "Knowledge comes from our senses, so if we extend our senses, we will consequently extend our knowledge" [18]. However, it is important to acknowledge that this is not necessarily intellectual knowledge that is being developed and later engaged.

Ernesto Neto, who develops interactive installations, has given us an example of how this can work practically. He said that his work is an experience of smell, colour, emotion, and sensory language, and that we "Receive information, but I want to stop thinking (…) I think that not thinking is healthy; it's like breathing life itself" [15]. As humans, we often have this need to stop thinking because our thoughts are an assortment of words, without leaving any space to feelings. Making drives us to that moment of abstraction and lets all those languages without words gain expression. The Fourthland project, whose work explores forgotten practices of social and environmental consciousness, says on their website about making: "it is the intangible heritage of humanity. Art practice and making with the hands can inform and initiate a healing of ourselves, the community and beyond" [17].

Taking the unwritten languages analysed as metaphors of human development and interaction, and understanding that humans have found different ways to express themselves when the possibility to translate everything into words does not exist, we can see that craftsmanship is a central concept for all of them. When David Brett talks about this topic in his book Rethinking Decoration [2], he defines the poetic of craftsmanship:

> This aspect is concerned with the making of it, with the fashioning of materials, and with the pleasure to be got from our encounter with the stuff of the world; and how all these come together to form a realm of meaning of its own kind that owes little or nothing to other domains.

Craftsmanship is based on making. During the making process, we embody the connection with materials, the acquired knowledge, and narratives. This action is what leads the maker to get engaged with basic human needs and become a medium between society and the products they make; by doing so, the maker is shaping the self but also having an impact on society. That is why making is always a social activity, because even unconsciously, it is always connecting embodied memories, materials, and culture. Everything that becomes social is deeply related to empathy, as empathy is what allows humans to get those longed for connections. Daniel Goleman [5, pp. 95–96] in his book *Emotional Intelligence: Why it can Matter more than IQ* highlights the importance of emotions to develop empathy and so be able to understand others better: "people's emotions are rarely put into words; just as the mode of rational minds is words, the mode of the emotions are not verbal".

6 How Can Ancestral Cultures and Unwritten Languages Inspire the New Concept of Luxury?

Not that long ago luxury was reserved only for people who could afford to buy very expensive items just for the pleasure of having something, rather than because they actually needed it. Only the rich and famous had access to the world of luxuries. Nowadays, the game has changed and what luxury means has shifted. We cannot talk about luxury anymore without considering words like health, happiness, creativity, and well-being as a part of it. Stylus, a trend researcher focused on consumers, published in 2018 a book about the macro trend of *"The New Rules of Luxury"* that if we talk about luxury we now need to focus on personal optimisation, mindfulness, development of wisdom, and individualism, because those are the things shaping what luxury means nowadays.

Analysing what we consider "beautiful" is a way to better understand this new concept of luxury. Luxury is no longer connected so strongly to aesthetics, and is much more bound up with the quality of experiences, what Kant calls "subjective sensation" or "feeling" [2, p. 21] leading to a more holistic meaning. How we analyse aesthetics comes first from what society has told us about beauty, and secondly from our own experience, from what has a positive impact on us and what gives us pleasure. That society often viewed beauty as being linked to (if not synonymous with) aesthetics was an unfair measure for something so important. Refocusing on the quality of experiences, as we are now, was a necessary corrective, because that is the beauty that engages our senses and makes us feel happier. The advantage of this is that everybody in the whole world is always going in search of pleasure and fulfilment, and it is something that we do not need to encourage people to go for, because that drive resides already inside every human being.

All these points make us look back to ancestral cultures and realise that even though luxury in the past was not even a concept for those societies, certain objects, ceremonies, and making processes were a space of joy, a source of happiness, and a sense maker of belonging, which overlaps to an extent with what we nowadays consider luxury. They found it not by looking for luxury but in the search of self-improvement, while at the same time strengthening their bonds to the cultures from which they came. When Socrates talks about the creation of a "Luxury state" he points out that in all probability a city where everyone is satisfied in terms of food, clothing, and shelter could be fairly balanced, but such a city does not fully cover human necessities. Human desires and feelings of fulfilment go beyond such bare necessities. The new concept of luxury tries to incorporate these deeper needs in our search for what makes us human, encouraging us to define ourselves by trying to reach a spiritual and personal evolution. Despite the differences in the communities analysed here, we can see that the development of unwritten languages and craft techniques made them sense a feeling of time and community, and offered them a knowledge that they could carry everywhere, that they could pass to anyone and where they could "write" the most interesting narratives about who they were, what they felt, what they made, and how this defined their way of being in the world.

Although these cultural values were mostly open to the whole community, many did not have access for different reasons, mostly because of their social status. This limited access is something that our society is suffering from as well. Not many people are aware of the importance of these values for our evolution as a society, and not many of them understand that by reaching higher levels of well-being we are not just improving ourselves but also positively influencing our surroundings. Likewise, not many people are willing to spend the required time to go beyond basic needs and understand, learn and consciously engage with alternative languages beyond the written and spoken. For the people that are already aiming for this new luxury, it is hard to understand that many others do not want to face changes consciously that they do not want to adapt to or do something about big social and environmental issues, and this is perhaps one of the saddest parts of luxury: even if this new concept of luxury is opened up to more people than before, it is still available only to those that are willing to look for it. The new concept of luxury, as with the way luxury was understood in the past, has the characteristic of being available just for the few.

Beyond this, what we cannot avoid in order to make this new luxury something attainable for more people is the need to recognise the immense connection that people in the past had with universal matters. These ancestral communities understood nature and the human body in a way that we are not capable of anymore. Now, the concept of luxury has changed, moving beyond the mere acquisition of goods and towards holistic experiences, via means that are responsible, sustainable, knowledgeable, and connected more with our internal needs.

There are three main aspects in luxury nowadays:

- Elevation to luxury status through time-invested craftsmanship
- Self-exploration about human needs in a world more focused on sustainability
- Personal experiences transformed into luxuries.

Craftsmanship is one of the key points in the new concept of luxury. As Joseph Beuys did with felt, craftspeople are doing with almost any material nowadays. All those forgotten old items or everyday materials are the main resources for craftsmanship, provided they are in the hands of experts in working with those materials. Old items and natural materials already hold a story, telling us something about the past, about where they came from. Innovators use these objects and materials as a blank canvas, putting body and mind to work together to give them a contemporary purpose or meaning. They also believe that a small number of crafted objects are more precious than many mass-produced ones. In craft it is not just the final result that matters, but the making process in and of itself that transforms the object into a "luxury".

This gives materials broader possibilities in the hands of craftspeople. As the documentary "The Future is Handmade" advocates, if we consciously access this new knowledge structure built by environmental, social, and cultural awareness, it can generate the same sense of belonging that ancestral techniques offered to their societies in a world that is increasingly making a lot of us placeless. That is one of the most interesting aspects of the new concept of luxury. The memories that materials and experiences leave in our bodies are things that cannot be removed from us. An expensive item can just be stolen, but knowledge and experiences have the value that cannot be lost or taken.

If craftsmanship is one of the keys to luxury, knowledge is the key to craftsmanship. It requires mastery of technique, and mastering a technique requires both time and a desire to fight back against a society that demands instant solutions to problems and instant gratification for desires. This is the space that craftsmanship gives, a place where we familiarise ourselves with techniques, materials, stories, where it is possible to consider knowledge as something that can be engaged by all the senses and where the only thing that matters is to be able to perform a task beautiful and consciously. This act is what produces the effect of being present in the world, offering well-being through this feeling of being connected.

Craftsmanship is one of the strongest challenges to capitalism, which used to be the main driver of the old concept of luxury. Those who could afford these goods would buy them, but without much conscious thought. Mass production developed as a way of spreading this beyond the richest elites, but the lack of consideration among buyers remained. Craftsmanship goes against this, while at the same creating products that are often considered more valuable, albeit now in a lasting and more sustainable sense. Craftsmanship also offers its value to anyone; anybody can become a craftsperson and make their own luxury or learn about it to appreciate and experience it in the right way. There is a recognition in craftsmanship of the importance of educating people, and this motivates them to add to their knowledge, take their skills to another level, and go beyond the predictable, by emphasising that these creators are human, with individual minds behind every piece. That is one of the best reactions to a world where everything looks like it has already been made, and where everything looks like a copy of something else. This leads us to the second concept about self-exploration and the consciousness that underpins luxury.

This self-exploration goes mostly to the new luxury user. Knowledge is not about what to buy anymore, it is about what is beautiful to me, where do I want to go, what do I want to do, and who am I? That is something that money cannot buy, which is why now knowledge has become the new currency, a very personal one that leads to many different paths. This makes us understand that luxury has to be personalised. This personalisation is derived from educating the consumer, showing them what is good for them, for their lives and for their surroundings, how they can leave a beneficial footprint on the world, and how they can create their own identity through the items they use and the experiences they have. Currently, we have the advantage of having a new generation with strong self-confidence, open to understanding how the world is changing towards preserving what we have and doing the most with it. We no longer need new things, we need to re-signify and update what we already have. This evolution is also a form of devolution; devolution not only means to "return", but it is also a form of administrative decentralisation, granting a level of autonomy, autonomy to learn, to develop, to encourage minorities and small but significant means of production.

When we look for trends nowadays, we can find many different ways to be on trend and to express ourselves, which means that even though we can always find a very strong leading trend, we can always build up our own conception of it. We can make choices based on our own values, and this is where self-exploration starts. This self-exploration is not always related to just our desires and our own pursuit, sometimes we are driven to develop it by being forced to react to certain things that are occurring around us and drawing our attention.

For example, one of these things that we can no longer ignore is that the world is going sustainable; why? Because it is a need that has a direct impact on us, but goes beyond humans and has an impact on the world around us. Resources are being wasted, the climate emergency is finally being described as such, and anxiety is spreading, all signs that that the consumer-driven way of living is obsolete. As makers, we have the power to offer good quality and sustainable products but we also have the challenge to educate consumers to select what is going to be good in the long term for them, their community and the environment. Craftsmanship, helped by natural material and human engagement, is helping to set up the mentality we need for the future: offering values of durability, excellence, and the benefits of the circular economy, but also offering a space where it is possible to slow down and find joy. This is a whole package, brands nowadays need to offer not just a product but an experience, not just be sure about their identity but also reach out to and connect with their customers' priorities and necessities.

As we mentioned before, the new concept of luxury involves experiences in many ways. How can experiences be translated into luxury items? Luxury brands cannot avoid the impact of young people opening new topics of discussion about matters that were irrelevant not long ago, nor can they avoid the importance of generating stimuli in the consumer beyond the pleasure of just buying, and finally they have to understand that we cannot have luxury if it has to be made at someone else's expense. All this leads to the creation of new luxuries in the most social, ethical, and sustainable way possible. This new way of analysing luxury highlights the status of master craftspeople: we now care much more about what they are doing, we recognise their mastery, and we are ready to hear stories through their hands. Some of them talk about their techniques, others about forgotten communities, and others still might be scientists or engineers experimenting with new materials in an innovative and sustainable way. Now as never before we are interested in narratives, and that is a plus for people who have things to say. Considering narratives are not in terms of facts or convictions but about a disclosure of the true meaning of life experiences.

From the perspective of the consumer, they also have a role to play because their voice manifests through their actions and behaviours. In the crazy political and economic times, we are going through, we understand that nothing can be taken for granted, and even if we cannot change the world, we need to collaborate to give voice to the things that matter to us. We are going to invest our money supporting fair causes, like encouraging local shops, making environmentally conscious choices, or helping individuals or communities to improve their situations. We are no longer in the "throwaway" era, we care about the things we are consuming, and we are heading towards an age of repair. The world is too small for the amount of waste that we are producing every day, and this matter is not just a question of space, but about damage to the environment, which will ultimately have the effect of damaging humans' health. Why do we need new things if we have plenty of materials that have already been used but are still around? We can use them as raw materials, giving a second life to them. We already went through a time where almost everything was (and some still are) disposable, but now thanks to craftsmanship we can access long-lasting products and have the chance to repair them in future if needed. This new mindset, and the new priorities it engenders, helps us to realise that the more accessible something is, the greater the chance that there is an associated cost in future either to us, to the environment, or both.

To conclude, I want to tell a story that from my point of view encapsulates all the concepts discussed above. In June 2019, a group of students from a school in Alta Gracia, Argentina did a research project to understand how clothes were made. They came across a specific loom that was used by the people who have lived in the area long ago to weave their own garments. They recreated this loom and started to weave with it. Looking to share their knowledge with the community, they organised an expo day in the main square of the town where the children not only demonstrated how to weave, but they also taught the community how to do so by themselves. Suddenly, an older man started to weave without any problem, and the woman next to him started to cry. No one understood the situation until she explained that the man, her husband, was suffering from Alzheimer's that he barely remembered anything at all, and this was the first time in a very long time that he had managed to retain something, namely the steps to do the weaving.

Experiences like this one are priceless. No one can put a price on the happiness of this wife seeing her husband acquiring and retaining new knowledge. No one could explain in words how this man understood the weaving process, but what we can do is to analyse the factors around this situation: we have the knowledge of the children that they obtained by getting engaged with the culture that was already present in the area but to a large extent forgotten. This knowledge was about mastering a manual technique that engages many of the senses. To take the technique and the knowledge to the next level, they decided to share it with the community, generating an awareness of the past of their own identity. After all these steps, the knowledge about the use of this loom has now been spread across the whole community. This made people interact and generate connections and these two factors generated well-being not just for the students or the elderly couple in question, but also for the people who witnessed the emotion of this scene.

Now, we are driven to feel emotions, to be more conscious about what is happening around us, and we cannot be blind about the role that we have to play. Bringing back the wisdom of ancestral languages and techniques and adapting them to our needs, we can now redefine the concept of luxury. It now is about capturing things and their whole meaning, like the Khipu system did. It is about offering pleasure and unforgettable memories through the engagement of the body in its entirety, considering humans as a whole entity with the need for connections in order to ensure well-being.

As Martien Van Zuilen says in the research *Through the Eye of a Needle*, textiles (but in this case, we could say craft products in general) can now be read like "a narrative text that chronicles and translates personal experience, cultural identification, and socio-cultural life embedded in multiple layers of meanings". The most meaningful narrative is no longer at the end, but in the process of making and discovery.

It is our task to encourage makers and craftspeople to keep going, to keep ancestral knowledge alive, to engage our senses more deeply, and to make us react to a turbulent world that needs people who can appreciate the value of staying connected in the most conscious way possible.

Bibliography

1. Aristotle (1972) De Memoria et Reminiscentia. In: Aristotle on Memory (ed and trans: Richard S). Duckworth, London
2. Brett D (2005) Rethinking decoration: pleasure and ideology in the visual arts. Cambridge University Press, Cambridge (Gran Bretaña)
3. Brett-Smith S (1982) Symbolic Blood: cloths for excised women. Rutgers University Community
4. Climo J, Cattell MG (2002) Social memory and history: anthropological perspectives. AltaMira Press, Walnut Creek, CA
5. Goleman D (2006) Emotional intelligence: why it can matter more than IQ. Bantam, New York
6. Heidegger M (1996) Being and time a translation of Sein und Zeit (trans: Joan S). State University of New York press
7. Kumar S (2013) Soil soul society: a new trinity for our time. Leaping Hare Press
8. Millar S (2008) Space and sense. Psychology Press, Hove, East Sussex, UK
9. Quilter J, Urton G (2002) Narrative threads, accounting and recounting in Andean Khipu. University of Texas Press, Austin
10. Steinter G (1991) Martin heidegger. University of Chicago Press; Reissue edition
11. Warin M, Dennis S (2005) Threads of Memory: reproducing the cypress tree through sensual consumption. J Intercultural Stud

On-line Resources

12. *Aluna Art Foundation (2016) Jorge Eduardo Eielson: on the other side of languages. Available at: http://www.alunartfoundation.com/jorge-eduardo-eielson-on-the-other-side-of-languages/. Accessed Aug 2019
13. Borer A (1996) The Essential Joseph Beuys. Thames & Hudson, London
14. *Frieze (2015) Fibre Is My Alphabet BY JENNIFER HIGGIE Jennifer Higgie talks to Sheila Hicks about the 60-year evolution of her artistic language. Available at: https://frieze.com/article/fibre-my-alphabet. Accessed Aug 2019
15. *Guggenheim Bilbao (2018) Ernesto Neto: the body that carries me. The Exhibition. Available at: http://ernestoneto.guggenheim-bilbao.eus/en/. Accessed Aug 2019
16. *Kabe contemporary (2014) Jorge Eielson: the other side of languages. Available at: https://www.kabecontemporary.com/jorge-eielson. Accessed Aug 2019
17. *Patreon (2017) Gladys Paulus textile artist is creating 'Hinterland'—an installation of ancestral healing costumes. Available at: https://www.patreon.com/GladysPaulusTextileArtist. Accessed Aug 2019
18. *TED Talks (2012) Neil Harbisson: i listen to colours (2012). Available at: https://www.ted.com/talks/neil_harbisson_i_listen_to_color? Accessed Aug 2019

The Artification of Luxury: How Art Can Affect Perceived Durability and Purchase Intention of Luxury Products

Matteo De Angelis, Cesare Amatulli, and Margherita Zaretti

Abstract Luxury brands are currently addressing the issues arising from the "democratization" of luxury consumption by looking for new ways to reinforce their aesthetic, moral and symbolic value. Along with this challenge, luxury brands are facing the growing consumers' concern about the social and environmental impact that luxury brands' activities bring forth. In this chapter, we propose that associating luxury products and brands with the concept of art and artworks might help luxury companies tackle these issues. We start from the definition of luxury and the analysis of the motives behind luxury consumption and then discuss the role played by sustainability in luxury through an overview of the main characteristics of luxury goods, such as scarcity and durability, that make them be considerable as sustainable in nature. Next, we discuss the idea that luxury and art share some important elements, such as the inherent strong emotional value, the relevance of craftmanship and savoir faire, and, above all, the idea of durability (defined as the ability of a product to maintain its quality and value over time), which characterizes both luxury products and artworks. In particular, building on this premise as well as on previous studies documenting the existence of the so-called *art infusion effect*—defined as the general positive effect that the presence of art in product advertising has on product evaluation and perception—we propose that the relevance of the artist's craftmanship in the process of the artwork creation positively influences consumers' perceived durability of the product advertised, which, in turn, positively affects consumers' purchase intention. Results of an experimental study discussed next support for our hypothesis. Theoretical contributions of our study and managerial implications of our findings are finally discussed in the chapter.

M. De Angelis (✉) · M. Zaretti
LUISS University, Viale Romania, 32 00197 Rome, Italy
e-mail: mdeangelis@luiss.it

M. Zaretti
e-mail: margherita.zaretti@studenti.luiss.it

C. Amatulli
University of Bari "Aldo Moro", Via Duomo, 259 74123 Taranto, Italy
e-mail: cesare.amatulli@uniba.it

© Springer Nature Singapore Pte Ltd. 2020
M. Á. Gardetti and I. Coste-Manière (eds.), *Sustainable Luxury and Craftsmanship*,
Environmental Footprints and Eco-design of Products and Processes,
https://doi.org/10.1007/978-981-15-3769-1_4

1 Defining Luxury

Several scholars, managers and actors operating in the luxury industry have attempted to provide a definition of what luxury is. However, the complexity of the luxury phenomenon is such to make the search for a unique and comprehensive definition a still open question [1, 2]. Indeed, as highlighted by Belk [3], the classification of a product or service as luxury is fluid and open to significant changes, since the perception of luxury is "specific to a particular time and place and is always socially constructed" (p. 41). For example, in the economically developed countries many products and services previously considered luxuries—such as mobile phones, foreign travel and computers—are today perceived as necessities [3]. However, while strong arguments exist about the fact that luxury products cannot be categorized as such by their appearance or intrinsic qualities [2], many contributions have been provided in the attempt to categorize luxuries and identify their peculiar characteristics.

Firstly, an open question is to what extent product expensiveness defines luxury. In economic terms, luxury brands are those which have constantly been able to justify a high price: a price significantly higher than the price of products with comparable characteristics [4]. Despite the prominence of price as a discriminant element of luxury products, this definition does not seem to be comprehensive enough to cover all features of luxury items, since it implies that the luxury status can be gained to goods having a price differential with other goods in the same category, included the upper-range brands [5].

A more comprehensive view of luxury has been provided by Ko et al. [6], who have highlighted three criteria to reach a thorough definition of luxuries: the existence of a sound conceptual foundation, the applicability to every product category and the measurability of the construct. Considering them all together, authors have agreed to theoretically define a luxury brand as "a branded product or service that consumers perceive to: be high quality; offer authentic value via desired benefits, whether functional or emotional; have a prestigious image within the market built on qualities such as artisanship, craftsmanship, or service quality; be worthy of commanding a premium price; and be capable of inspiring a deep connection, or resonance, with the consumer" (p. 406). Along this line, Vickers and Renand [2] argue that luxury goods can be "usefully defined in terms of a mix of components of functionalism, experientialism and symbolic interactionism" (p. 472), in which the role of social and individual (psychological) cues is significantly more relevant than in the context of non-luxury products. Mortelmans [7] has gone even further, defining luxury products as those "that have a sign-value on top of (or in substitution of) their functional or economical meaning" (p. 510) or, in other words, as objects to which the consumer society attaches additional and undetermined meanings that are independent from any functional and economical logic.

To conclude, although many authors have attempted to account for the multidimensionality of luxury, the fluidity and the complexity of the notion of luxury have made it difficult to come up with a comprehensive definition. A deeper discussion about the personal and interpersonal role played by luxury, whose distinctiveness lies in its psychological value [8], can thus help to clarify.

2 Luxury Consumption: Personal Versus Interpersonal Motives

According to Amatulli and Guido [9], luxury can be broadly categorized into *externalized* and *internalized*, on the basis of the idea that individuals purchase luxury goods under the influence of both interpersonal and personal motives. Externalized luxury is a socially embedded construct as it refers to individuals' tendency to purchase luxury under the influence of others with the aim to socially position themselves. Conversely, internalized luxury is an individually embedded construct as it refers to the purchase of luxury goods for the satisfaction of consumers' personal needs and tastes. In this case, consumers go "beyond externally imposed criteria" [8, p. 125] and are driven by personal lifestyle, emotions/hedonism and culture [9], as well as by the need to transform their identities into an ideal self and to find support for their own individual identity [10].

Such a dichotomy grounds its root in a number of theories about drivers of luxury consumption (see Table 1 for a summary of these theories). The oldest theory is that built off the idea of conspicuous consumption. In 1899, Veblen introduced this label to define individuals' tendency to consume goods in a high visible way in order to impress others and make them infer their wealth and power, particularly in the context of high socioeconomic mobility. Mauss [11] resumed Veblen's reflections to explore their application in the gift-giving social dynamics, arguing that conspicuous and costly generosity is the most powerful means to signal status. The author anticipated what would have been lately conceptualized as the signaling theory, which states that the individual's disposal of sufficient personal resources can be inferred by his or her engagement in generous or wasteful behaviors that, perhaps paradoxically, are too costly to be fake [12].

Defined as "the behavioral tendency to value status and acquire and consume products that provide status to the individual" [14, p. 34], status consumption is strictly related to, yet still separate from, conspicuous consumption. In other words, status consumption relies on the desire to enhance social status by owning luxury goods (which may or may not be publicly displayed), while conspicuous consumption focuses more on the overt display of luxury and wealth. Building on extant studies, Eastman and Eastman [19] focused on the identification of internal and external motivations for status consumption. External motivations are interpersonal and extrinsic, focusing on the social effects that the ownership of luxury goods produces. They include the above-mentioned need for conspicuous consumption, exclusivity (snob luxury purchase behavior) and social identity (bandwagon luxury purchase behavior). Conversely, internal motivations are personal and intrinsic, focusing on the expression of inner values and tastes. They include hedonism, self-reward and

Table 1 Most relevant theories explaining luxury consumption

Interpersonal motives		
Conspicuous consumption	Individuals' tendency to consume goods in a high visible way in order to impress others and make them infer a condition of wealth and power	Veblen [13]
Status consumption	Behavioral tendency to value status and acquire and consume products that provide status to the individual	O'Cass and McEwen [14]
Social comparison	Individuals' tendency to evaluate themselves through objective and nonsocial means that, if not available, will be replaced by the comparison to other people	Festinger [15]
Uniqueness	Individuals' tendency to engage in nonconformist behaviors and to differentiate themselves when the degree of similarity relative to others is perceived as excessively high	Snyder and Fromkin [16]
Personal motives		
Self-concept	Individuals' cognitive and emotional perception of themselves	Rosenberg [17]
Extended self	Individuals' tendency to consider their possession as part of their self-concept	Belk [18]

perfectionism. Indeed, status consumption can be motivated by the search for emotional and sensorial gratification, by the desire to reinforce the self-concept through self-reward and by the search for high quality.

Another relevant contribution has been provided by social comparison theory, according to which individuals tend to evaluate themselves through objective and nonsocial means that, if not available, will be replaced by the comparison to other people [15]. Since this theory also states that individuals tend to conform to the prevalent opinion of the social groups they belong to, Wiedemann et al. [20] have argued that luxury goods, by enclosing prestigious values, may be used to conform to social standards.

According to self-concept theory, moreover, self-concept is defined as the "totality of the individual's thoughts and feelings having reference to himself as an object" [17, p. 7], i.e., how the individual cognitively and emotionally perceives himself or herself. Since general agreement exists regarding the expressive and symbolic value of luxury—defined as the ability of the product to convey a psychological meaning [21]—self-concept can be a motivator for luxury consumption. Moreover, Kastanakis and Balabanis [22] have provided evidences of the fact that differences in self-concept orientation can lead to differences in luxury consumption: For example, bandwagon luxury consumption is more likely to occur among consumers with an interdependent self-concept than among those with an independent self-concept.

The concept of extended self can be extremely useful to understand luxury consumption motivations as well. Conceptualized for the first time by Belk [18], the extended self refers to the phenomenon according to which individuals define themselves not only through what it is seen as "me," but also through what is seen as "mine." Specifically, individuals perceive their possessions as part of their self-concept and use them to get closer to their ideal self that is to whom they hope to be. In this perspective, the luxury symbolic value previously mentioned may play a fundamental role in making luxury consumption an effective tool for self-extension.

Last is the theory of uniqueness. Developed by Snyder and Fromkin [16], it states that although people sometimes feel the need to conform, they may tend to engage in nonconformist behaviors and to differentiate themselves when the degree of similarity relative to others is perceived as excessively high. This need for uniqueness can be fulfilled by the consumption of luxury goods, which are by definition scarce and rare [23].

3 The Luxury Dream and Scarcity

The basic law of economics states that when demand exceeds supply, price increases. This imbalance is the *sine qua non* condition for luxury to exist, since scarcity represents the core of luxury brands' DNA. Indeed, luxury goods need to be admired by all and owned by few to be properly considered as such [5]. Their power relies on their magical aura of unattainability, which is psychologically evoked by their incorporation in the lifestyle of extraordinary people and practically fostered by physical rarity. As argued by Dubois and Paternault [24], the deeper the gap between awareness and penetration of the luxury brand the stronger its desirability, since "awareness feeds the dream but purchase makes the dream come true and therefore contributes to destroy it" (p. 73).

Without prejudice to the general validity of what mentioned above, Kapferer [25] offers interesting insights regarding the concept of rarity and its different applications in the luxury context. The first and most intuitive type of rarity is the objective rarity that is the outcome of the material limitation imposed on the production and commercialization of luxury goods, achieved through high prices and limited production. As an example, Lamborghini's CEO Stefano Domenicali has recently stated that no more than 8000 vehicles per year will be sold from 2020 onward [26], demonstrating the company will reinforce its exclusivity. However, it is worth mentioning that physical scarcity clearly places a limit to luxury brands' growth and consequently risks compromising both company's well-being and shareholders' expectations. Therefore, a shift in what the author defines "virtual rarity" is currently taking place in the luxury sector. It represents an ephemeral and artificially induced type of rarity, which makes the perceptions of exclusivity and privilege arise without sacrificing sales. The means through this type of rarity achieved are many: the regular launch of limited editions, the adoption of an exclusive distribution strategy and the careful selection of the messages to be communicated. As an example, Louis Vuitton has recently partnered with six of the world's most renowned artists to create the limited edition of Artycapucines collection [27].

4 Luxury and Sustainability

Defined as the "development that meets the needs of the present without compromising the ability of future generations to meet their own needs" [28], sustainable development is today a very urgent issue concerning both individuals and companies. The urgency to find practical solutions to assure a sustainable growth has been fueled by the rise of a throwaway society that "relies on the relentless production of novelty by firms and the relentless consumption of novelty by households" [29, p. XVI]. Keen sociological insights regarding this phenomenon have been provided by Bauman [30], who has highlighted the liquidity of postmodern times: In a society obsessed by appearance and individualism, individuals anxiously look for immediate gratification of their constantly evolving temporary identities, distancing themselves from the immobility imposed by traditional and heavy objects of consumption and embracing the liquefaction of goods. Solid accumulation is replaced by elimination and substitution that gratify individuals' need for fluidity. This trend has been both powered and exploited by product suppliers, who often take advantage of planned obsolescence. The concept of planned obsolescence consists in producing goods which will be "no longer functional or desirable after a predetermined period" [29, p. 4], by cutting production costs at the expense of quality and encouraging consumers to replace the product (sometimes leveraging on technological development).

Unfortunately, such a sociological and economic context does not allow to achieve a really sustainable development. Importantly, Elkington's "triple bottom line" model, proposed in 1994, identifies three main dimensions that should be taken into consideration by those companies that desire to pursue a sustainable development. The economic dimension refers to the ability of the company to generate enough profits to ensure business and technological development, as well as employees' stability over time. The ecological dimension refers to the ability of the company to minimize its environmental impact by optimizing its production processes. Finally, the social dimension refers to the commitment of the company to improving collective well-being and promoting the cultural heritage.

In light of the existence, according to the bulk of extant scientific research, of a sort of contradiction between the values associated with luxury and those associated with sustainability (see [31] for a review on this topic), whether the luxury sector can be considered sustainable or not is a highly debated issue. Indeed, in a surely too simplistic way, sustainability has been typically associated with concepts such as altruism, sobriety and morality, while luxury has been typically associated with excess, ostentation and superficiality [32]. In contrast, however, several authors believe that luxury is among the sectors that best fit with the conditions enabling sustainable development and that most distance themselves from the values of the throwaway society [33, 34]. To illustrate, according to Kapferer [35], luxury and sustainable development converge because they both focus on rarity and durability. The objective rarity that fuels luxury value characterizes both materials and craftmanship. On the one side, resources are precious and require to be protected to guarantee the future of the sector itself. On the other side, the crafted nature of luxury products requires the involvement of rare savoir faire and fine artisans' skills, in contrast to the unskilled labor exploitation that oftentimes characterizes the mass fashion industry. Durability is a key element too: Since they typically never go "out of fashion," luxury products are conceived and produced in a way that preserves their aesthetic and functional value as long as possible.

Amatulli et al. [31] have proposed a quite detailed analysis of the reasons why luxury items might be *inherently* sustainable, possibly more so than their mass-market counterparts. The first reason lies in luxury goods' high quality: By ensuring materials' safety and reliability, they certainly provide positive benefits to consumers. The second reason lies in durability and rarity: Luxury products are by definition scarce, and their longer life span prevents frequent purchase of alternative and more perishable products, with positive outcomes in terms of resource outflow and waste. The third reason lies in craftmanship: The employment of talented, highly skilled and experienced labor force by luxury firms produces a positive social impact and ensures the preservation of traditions and heritage. To conclude, luxury goods stand for sustainable, long-term investments that are consistent with the concept of "circular economy," which states the necessity to extend products' life within the economic and social system in order to minimize waste and make the most out of material resources.

5 The Growth of Luxury Market

Except for the years encompassing the last financial crisis, occurred between 2007 and 2009, the luxury market never stopped growing in the last thirty years [36]. The overall value of luxury market has been estimated at about € 920 billion in 2018, and a 4–5% annual growth is expected until 2025 [37]. Valued € 330 billion in 2018, the personal luxury sector will enjoy a 3% annual growth until 2025, driven by accessories and cosmetics. The experiential luxury sector, valued € 590 billion in 2018, is supposed to grow even faster, enjoying a 5% annual growth until 2025 [37]. Data clearly shows that the luxury sector has never been as prosperous as it is today. Its success appears to be driven by two main forces. On the one hand, the impressive growth of digital channels, that cover today around 10% of total sales in the personal luxury sector, while on other other hand, the impressive increase of luxury consumption among Chinese consumers, that today account for about 33% of global luxury purchases [36].

It might be certainly interesting to reflect on the reasons underlying the luxury business' growth. As highlighted by Kapferer [38], in the past luxury was a prerogative of the *happy few*, namely a small circle of powerful individuals who used to leverage on luxury consumption in order to show their taste and impress crowds [39]. In the nineteenth century, the rapid rise of a new middle class has led to the fast expansion of luxury customer base and to the gradual "democratization" of luxury desire. As illustrated by McNeil and Riello [40], this phenomenon, firstly emerged in Europe, has later happened in several emerging countries, which entered into the luxury market just at the time when saturation was being reached. In particular, in the 2000s China has gone from being a "virgin land for luxury" [40, p. 245] to being the motherland of those consumers "who have made the fortune of those European luxury brands that today account for the lion's share of the market" [40, p. 247]. Indeed, luxury brands started having a great appeal especially to young middle-class Chinese consumers, "who have no personal recollection of China under the duress of strict communism" [40, p. 247]. Strong growth in the middle class has also characterized BRIC countries [40]. In particular, luxury has recently become the symbol of wealth for those Russian who have "accumulated enormous fortunes since the fall of communism in the early 1990s" [40, p. 250], while the rise of Brazilian middle class has made the country the second largest market for several luxury services [41].

6   The Challenges Posed by Luxury Growth and the Resolutive Role of Art

While growth and product diffusion are not an issue for premium brands, which indeed benefit from the increase of their market share, they conversely may represent an obstacle for luxury brands' well-being. In fact, besides the unquestionable short-term economic benefit, sales growth can also cause the exposure of luxury firms to the risk of losing their exclusive appeal and their narrative meaning [42]. In other words, the extension of luxury customer base from the *happy few* (the powerful élite) to the *happy many* (the wealthy middle class), thus the transformation of luxury "from the ordinary of the extraordinary people to the extraordinary of the ordinary people" is likely to distance the élite itself, which is populated by those individuals who "ensure the long-term desirability of the brand" [38, p. 373]. Moreover, the expansion of luxury has created a condition of overexposure of luxury brands, which have invaded individuals' everyday life by massively resorting to generic media channels such as television, newspapers, magazines and Internet [40]. This phenomenon has exacerbated the moral criticisms that always existed around luxury, blamed to foster social inequality. Such arguments call into question the reputation of the luxury sector and even its right to exist [38].

Aware of the above-mentioned issues, Kapferer [25, 38] claimed the centrality of art in the resolution of the problems posed by the growth of luxury. Able to provide a powerful aesthetic and moral endorsement, art can strengthen luxury symbolical authority and offer new segmentation criteria beyond price (such as cultural and humanistic sensitivity), establishing a dialogue with a new, creative and post-materialistic élite populated by those "extraordinary people" who are destined to shape the future. The involvement of art, therefore, not only guarantees the loyalty of the *happy few*, but also defuses the moral criticisms related to luxury consumption by taking it to a higher and unquestionable level of meaning. Moreover, associating luxury with art may also help discouraging new potential competitors from entering the luxury market. Indeed, competitors may have difficulty decoding and imitating this source of competitive advantage, since the art–business collaborations are usually complex to define, highly specific and tacit [43]. Kastner [44] identified three benefits stemming from the associations between art and luxury brands. Firstly, this association "entrenches in the consumer's mind the perceptions of originality, ingenuity and inventiveness" (p. 45) and enshrines the aesthetic sensitivity of the brand. Secondly, it enriches brand content and invigorates the aspirational storytelling typical of luxury brands. Thirdly, the art–luxury association elevates luxury brands above the purely commercial dimension.

As reported by Kastner [44], the relationship between art and luxury experienced relevant evolutions over the twentieth century. In the 1930s, the surrealist painters Salvador Dalì and Jean Cocteau partnered with the Italian fashion designer Elsa Schiaparelli in order to bring their creativity in her clothing designs [45], while the Italian shoe manufacturer Ferragamo launched an advertising campaign realized in collaboration with the futurist painter Lucio Venna [46]. In the 1960s, Yves Saint Laurent launched its haute couture collection taking explicit inspiration from Piet Mondrian's artworks [47]. In the 1980s, longer-term forms of collaboration were introduced, such as the creation of the art museum Foundation Cartier in Paris [48]. Finally, the huge commercial success of the high-profile artistic collaborations initiated by Louis Vuitton's artistic director Marco Jacobs in the late 1990s made this phenomenon definitely popular among luxury brands.

Today, *artification*—defined as the process in which "something that is not regarded as art in the traditional sense of the word is changed into something art-like or into something that takes influences from artistic ways of thinking and acting" [49]—involves several short-term and long-term elements of luxury brand management. Chailan [43] has provided an overview of the main long-term activities that luxury brands undertake to link themselves with art. Taking into consideration both the intensity of the engagement and the time perspective of the relationship, the author has identified four types of links: artistic mentoring, artistic collaboration, foundations and patronage. Given their long-term nature, all these strategies fall outside the ordinary management of luxury brands and thus represent exceptional activities which require both managerial and financial extra-efforts. But also the most ordinary elements of the luxury marketing mix—such as retailing, merchandising and advertising—are actively involved in the process of *artification*. In this respect, as highlighted by Joy et al. [50] through an ethnographic study about consumers' experience in Louis Vuitton flagship stores, luxury retailing is rapidly changing by turning stores into hybrid institutions embodying elements of both museums and art galleries. This is achieved by leveraging interior design, effective lighting, artisanal merchandising and suggestive product display, all managed together with curatorial attention. Advertising is another element of the marketing mix of luxury brands which is deeply involved in the process of *artification*. The effect that the introduction of artistic elements in product advertisement has on product perception and evaluation—named the *art infusion effect*—has been extensively explored by several scholars. This subject, which is at the core of the present study, is discussed in the next section.

7 The "Art Infusion Effect"

Despite the general awareness that marketers and scholars have shown regarding the positive effect that the introduction of artistic elements can have on consumer evaluations [51], neither empirical evidence nor theoretical conceptualization existed until Hagtvedt and Patrick's studies [52]. They proved the existence of the *art infusion effect*, which is the general and positive "influence of the presence of art on consumer perceptions and evaluations of products with which it is associated" (p. 379). Authors theorize that it represents a special kind of spillover effect, since the "perceptions of luxury associated with visual art spill over from the artwork onto products with which it is associated, leading to more favourable evaluations of these products" (p. 379). The spillover effect—namely, the transfer of properties from one object to another, linked to the previous one [53]—is the impact that information provided in product-related messages has on beliefs about attributes that are not mentioned in the messages [54] and has proved to be more influential in terms of attitude changes than the message itself [55]. The effect is explained by consumers' tendency to infer missing attribute information relying on their intuitive notions of inter-attribute correlations and on the correlational information gathered from the message [54].

Hagtvedt and Patrick's [52] findings have been later challenged by further studies, which expanded the above-mentioned theoretical framework by bringing in the mediating role of emotions [56]. In particular, authors supported prior findings regarding the mediating role of luxury perception and also demonstrated across three studies the mediating role of brand affect—namely the positive emotional response to the brand [57], arguing that emotion represents "the lower-level mechanism driving the higher-level effect of perceived luxury on product evaluations" (p. 403). Results fully supported assumptions: when introduced as additional mediator, brand affect significantly and fully mediated the *art infusion effect*, while perceived luxury became nonsignificant.

7.1　How Product Type and Price Affect the Art Infusion Effect

Since artwork can be potentially included in the communication of any type of product, several scholars have tried to assess if any differences in the art infusion mechanism exist among different product types. The most significant distinction among products is drawn taking into consideration the needs they are aimed to satisfy. Hedonic goods are the ones whose consumption is sensorial and both affectively and aesthetically driven, being linked to sensual pleasure, fantasy and fun [58]. Conversely, utilitarian goods are the ones whose consumption is cognitively driven, goal-oriented and linked to the completion of a functional task [59]. Although products can embody both hedonic and utilitarian dimensions to varying degrees, a primary classification of them among the two categories can usually be drawn by consumers [60].

Whereas evidences about the effectiveness of the *art infusion effect* among hedonic and utilitarian goods diverge, most of the studies have demonstrated that the use of art better fits with hedonic products. Taking inspiration from the semantic matching process theorized by Collins and Loftus [61], Huettl and Gierl [62] claimed that a positive *art infusion effect* on luxury perception exists only for hedonic products. Hüttl-Maack [63] further extended existing knowledge by testing whether ambiguous and only moderately hedonic product differs from hedonic ones in terms of art infusion efficacy. Evidences show that, while in the highly hedonic condition "uniformly beneficial effect of art are found" (Hüttl-Maack, p. 271), in the more ambiguous condition the effect of art's presence is influenced by consumer's interest in art. In particular, the presence of art has a positive effect among art-interested consumers and no effect on non-interested ones: While the former individuals "are motivated to engage in processing to make sense of this combination," the latter ones "seem to be not motivated for this step" (Hüttl-Maack, p. 271).

As mentioned before, moderate disagreement exists among studies. Estes et al. [56], who investigated the mediating role of emotions in the *art infusion effect*, counterintuitively stated that the presence of art increases brand affect more for utilitarian products than for hedonic ones, due to a phenomenon of diminishing return. Since "hedonic products primarily possess emotional attributes, the additional affect induced by an artwork has a smaller influence on evaluations. And contrarily, because utilitarian products tend to possess functional attributes, increasing affect via art has a relatively large effect on evaluations" [56, p. 404]. To conclude, product type has proved to significantly influence the art infusion mechanism. Despite the lack of full agreement on the subject, the bulk of scientific research seems to confirm that the *art infusion effect* is more likely to be successful for among hedonic goods.

Some scholars have also investigated the role that price difference and price information play in the *art infusion effect* and, in particular, on perceived prestige of the advertised product. Lee et al. [53], who investigated the interaction effect of price differences and art's presence on perceived prestige, demonstrated that the presence of art in the product advertisement has a positive effect on perceived prestige, regardless of the product price. Conversely, in the non-art condition a moderately higher price difference has proved to affect perceived prestige more than a non-price difference strategy. Indeed, while the presence of art itself enhances perceived prestige and weakens the effect of price difference, the absence of art makes the price more relevant, given its role in signaling customer's status, indicating product quality and enhancing perceived product exclusivity. Surprisingly, authors have also observed that, in the non-art condition, price difference has no effect on perceived prestige when too high (i.e., 150% price difference). Despite the lack of full clarity on the non-art condition, the research definitely confirms that "products could benefit from artwork to increase their price above their regular price" (Lee et al. [53, p. 602].

Some scholars have also questioned whether the positive effect elicited by the association of art with a product can be undermined by the increase of perceived expensiveness that is likely to be aroused by this association and whether the mentioned interference mechanism is affected by price information (e.g., [62]). Indeed, artwork scarcity and uniqueness can elicit perceptions of expensiveness that may spill over onto the product in the same way luxury perceptions do. However, this effect can be mitigated or even neutralized by providing information about the product price within the message. Findings have supported authors' hypothesis, proving the coexistence of interfering effects caused by the spillover of both luxury and expensiveness perceptions and the inhibiting role that price information plays on the negative expensiveness-related effect.

7.2 Beyond Product Evaluation: How Art Affects Consumers' WTP and Brand Extendibility

Different authors contributed to better understand what the presence of art in product advertising can affect, exploring other variables besides product evaluation and perception. Hüttl-Maack [63] contributed to this goal by giving evidences of the fact that, besides product evaluation and perception, the use of art in product communication is also able to affect consumers' willingness to pay (WTP), being the former a key determinant of the latter [64]. This research not only "extends the prior findings conceptually by showing that not only expensiveness but also the perceived value behind it is increased" (p. 271), but also highlights how art can affect consumer's behavioral response, being the WTP a much closer measure of it compared to product evaluation.

Other relevant insights have been provided by Hagtvedt and Patrick [65], who further expanded the knowledge related to the *art infusion effect* by demonstrating how it can be used to enhance brand extendibility. As recalled by the authors, prior research about brand extendibility has identified two main factors affecting it, namely the parent brand image/quality and the category and conceptual fit between the new product and the parent brand [66, 67]. They argued and demonstrated that both constructs can be positively influenced by the presence of art, leading to more favorable evaluations of brand extension. On the one hand, the presence of art enhances brand image through the spillover of luxury perception. On the other hand, it increases the perceptions of category and conceptual fit by intensifying consumer's cognitive flexibility: Improving consumers' capacity to integrate information in non-obvious ways, art enables consumers to draw meaningful patterns among stimuli, even if divergent.

7.3   The Content-Independent Influence of Visual Art

Along with the original demonstration of the *art infusion effect* existence, Hagtvedt and Patrick [52] provided evidence about its content-independent nature. In other words, they showed that the positive influence of art on product evaluations and perceptions does not depend on the content of the artwork used to advertise the product, but rather "on general connotations of luxury associated with visual art" (p. 379). If it did, the *art infusion effect* would not be generalizable, since the valence of product evaluation would depend on the positive or negative valence of the artwork content. The independence of the *art infusion effect* from artwork content is explained by the fact that art has "general connotations that are positive per se" (p. 381) and that make it intrinsically different from any other sensory phenomena that, in turn, can have positive or negative valence (such as smell, sound and non-art visual stimuli).

Neural responses to art images have also been investigated in order to test the content-independent nature of the *art infusion effect*. By performing an event-related functional magnetic resonance imaging (fMRI) study, Lacey et al. [68] demonstrated that the exposure to art images, regardless of their content and style, activates regions of the brain (i.e., ventral striatum, hypothalamus and orbitofrontal cortex) related to reward circuitry, whose key function is to drive behavior and decision making under conditions of uncertainty [69]. The same regions did not appear to be engaged in the presence of non-art images.

Deepening the comprehension of the role of the art content in the *art infusion effect*, Hagtvedt and Patrick [70] investigated across three studies the distinction between art content and manner, providing evidence of the fact that, while "artwork as art" is context-independent, "artwork as illustration" is context-dependent. Content and manner represent two components of the artwork: While the former concerns "what is depicted" and conveys information, the latter concerns "how it is depicted" and distinguishes artworks from mere illustrations. Findings showed that, when the artwork content is not stressed, it does not affect product evaluation. Conversely, when the artwork is no longer perceived as art because of the shift of emphasis from manner to content, product evaluation is affected by the fit between the artwork content and the product. Moreover, the third study conducted by the authors demonstrated that consumers' mind-set can drive them to process the artwork as either art or mere illustration. Since abstract mind-sets are "associated with higher-level construals and schematic, global processing" (p. 1628), they are more focused on the image category (i.e., visual art) and thus will be unaffected by artwork content. Conversely, since concrete mind-sets are "associated with lower-level construals and attribute-level, local processing with a focus on contextualized features" (p. 1628), they are more focused on particulars of the image and will be affected by artwork content.

To conclude, we can state that the *art infusion effect* is content-independent. However, the content of the artwork can become relevant if consumers' attention gets focused on it. Indeed, in this case the artwork will be no longer processed as belonging to the art category, being perceived as a mere illustration.

7.4   *The Role of Art Saliency in the* Art Infusion Effect

Some authors have investigated how product and brand evaluations and perceptions can be affected by two possible expressions of art saliency, namely artwork recognizability and artist's iconicity. Peluso et al. [71] examined the role that art recognizability plays in the art infusion mechanism and how it is affected by consumers' characteristics and needs. Recognizable artworks are the ones that can be easily identified by most individuals. Embodying the most recurrent colors, techniques and images used by the artist, they are able to "elicit the artist's image in people's minds" [71, p. 2194] and to conform to preconceived mental schemas [72]. Authors provided evidence of the fact that these artworks, compared with less recognizable ones by the same artists, increase the luxury perception of products advertised together with them, thus their evaluation. Indeed, being highly recognizable artworks usually perceived as more valuable and sold at a higher price than the less recognizable ones by the same artist, they are able to infuse a higher sense of prestige and exclusiveness that spills over onto the related product.

As mentioned, authors also questioned if the needs supposed to be main drivers of luxury consumption—namely the desire to signal status and the desire of distinction—influence the above-described findings. A consistent part of luxury consumers purchase luxury products that display easily recognizable luxury qualities because they use luxury consumption as means to demonstrate their higher social standing and wealth to others. Authors demonstrated that these consumers, compared to those who show a lower desire to signal status, are "more interested in buying luxury products promoted in advertisements that feature recognizable artworks rather than non-recognizable artworks" [71, p. 2195]. Conversely, other consumers purchase luxury products to differentiate themselves from the other luxury consumers and to convey their uniqueness. These consumers generally use uncommon signals of luxuriousness and thus prefer rare, special or nonconforming luxury products. Authors proved that these consumers, compared to those who are characterized by a lower desire for distinction, are "more interested in buying luxury products promoted in advertisements featuring non-recognizable artworks rather than recognizable artworks" (p 2196). To conclude, the positive effect of art recognizability on the art infusion mechanism has been demonstrated. Furthermore, it has been shown that consumer needs can change it: While the evaluations of consumers who aim at signaling their status are more positively influenced by highly recognizable artworks, evaluations of those who aim at differentiating from others are more positively influenced by hardly recognizable artworks.

Another issue that the literature has explored is how the image of the artist who produced the art associated with the product advertised can affect product evaluation and perception. Specifically, Scarpaci et al. [73] investigated the effects that artist's iconicity has on the perception of national brands. Artists, who "may be viewed as uniquely positioned representatives of their culture" (p. 321), often serve as national icons by embodying national identity and values. Through a qualitative study, these authors demonstrated that artists' association with a national brand enables an "icon myth transfer effect" that is a mechanism of translation of cultural and national values into brand associations [73]. To do that, they distanced from the main characteristics of previous studies: Taking into consideration two forms of artistic expression (dance and poetry) that do not have a visual display, they introduced the artistic element into the experiment by considering two products named after famous artists. This research has contributed to both enrich the knowledge of the *art infusion effect* and extend the boundaries of cultural branding. On the one hand, it demonstrated that the association of the product with art can generate positive effects because of the spillover not only of luxury perception, but also of cultural values. On the other hand, it challenged the most traditional views of cultural branding, in which "the brand is the myth, and the myth reflects characteristics of the myth market" [73, p. 330].

8 A New Theoretical Model

In the present work, we introduce a novel conceptual framework that aims to advance extant knowledge about the art infusion effect in two important ways. First, we introduce the role of a construct that, while being peculiar to both art and luxury, has not been investigated by previous research in this stream: the perceived durability of the product advertised. Second, we introduce a distinction between artworks based on what we call *intensity of artist's craftmanship,* which we predict to affect consumers' perceptions of product durability and, in turn, purchase intention (see Fig. 1).

Indeed, some scholars have highlighted that the luxury sector can be intrinsically sustainable (e.g., [31]), and more specifically, two luxury goods' peculiar characteristics, such as craftmanship and durability, are particularly important to respond to stakeholders' growing concern for the environmental and social impact of companies' activities. Craftmanship—namely the skill with which something is made by hand—is not only "synonymous with time and the specialized labor needed to produce an object of value, a symbol of tradition passed down from generation to generation, the fruit of manual know-how", but also a "type of guarantee in terms of quality, duration and aesthetics" [8, p. 130]. Thus, consumers link the intensity of craftmanship to product durability, which literature considers as one of the main drivers of luxury value perception [74]. Building on knowledge about *art infusion*

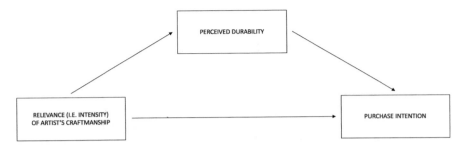

Fig. 1 Effect of intensity of artist's craftmanship on purchase intention through perceived durability

effect and its underlying spillover mechanism, we therefore argue that, in the context in which the product is advertised through the use of artistic elements, the intensity of artist's craftmanship in the process of artwork creation may affect the perceived durability of the product, since the savoir faire elicited by the artwork may spill over onto the product itself. We also argue that, in turn, perceived durability may affect consumers' purchase intention, being it considered a relevant driver of luxury purchase. Specifically:

> *H1. When a luxury product is advertised through an artwork, the intensity of artist's craftmanship in the process of artistic creation positively affects perceived product durability.*
> *H2. When a luxury product is advertised through an artwork, perceived product durability mediates the relationship between the intensities of artist's craftmanship in the process of artistic creation and consumer purchase intention.*

9   Empirical Study

In order to empirically test our formal hypotheses, we conducted a quantitative study aimed at demonstrating that the intensity of artist's craftmanship positively influences product durability, which in turn increases consumer purchase intention. Specifically, we conducted an experiment based on a two-condition between-subject design. Data has been collected through an online pull of prepaid respondents and analyzed using SPSS statistical software.

9.1   *Procedure and Sample*

The survey, created through the Qualtrics platform, has been distributed online through different social media channels. Being destined to be completed by Italian respondents, it has been fully drafted in Italian. Once 270 observations have been collected, the questionnaire has been closed and data cleaned. Specifically, all the incomplete observations as well as the ones which have failed the attention check have been deleted in order to improve data quality and reliability, leading to a reduced sample of 215 respondents. The final sample was equally distributed by gender (51.2% female and 48.8% male) and composed of 39.1% students, 14.4% independent contractors, 34.0% employees, 9.8% executives and 1.8% pensioners or unemployed. Most of the respondents (61.9%) were aged between 18 and 34, while 20.4% were aged between 35 and 54, and finally 17.7% were 55 years old or older. The majority of respondents, about 77% of the sample, declared to have a bachelor's degree or a higher academic qualification.

At the beginning of the questionnaire, respondents' attentiveness has been solicited by asking them to pay particular attention to the image and text they would have seen immediately after. Moreover, they have been informed that the following were fictional scenarios. Respondents were then randomly assigned to one of the two scenarios resulting from the manipulation of the independent variable, namely the intensity of craftsmanship involved in the creation process (102 respondents were exposed to the low-intensity condition, and 113 respondents were exposed to the high-intensity condition). In order to avoid any potential influence coming from respondents' previous experience with the brand, we decided to resort to a fictional one created for the purpose that is Kéntro. In both scenarios, it has been said to respondents that the brand Kéntro has decided to promote its new wallet collection by incorporating in its visual advertising the artwork of a famous artist. In the low-intensity condition, the artist was Lucio Fontana, a major abstract artist. In the high-intensity condition, the artist was Sandro Botticelli, a leading figure of Renaissance art. Moreover, we provided respondents with a brief overview about the art genre the displayed artwork was representative of, stressing the level of importance of the artist's craftsmanship (low in the abstract condition and high in the Renaissance condition). The above-described text has been paired with a figure showing the Kéntro's advertising. In both conditions, the visual stimuli have been realized showing the brand's logo, the artwork and the product advertised, on a white background (see Fig. 1). We decided to use a wallet in our stimuli because this product category is widely used by the majority of people in everyday life, thus to reduce the risk that the purchase intention could be affected by respondents' potentially low usage. Moreover, accessories seem to represent the most relevant product category in the luxury fashion sector.[1] In addition, to avoid a potential effect of respondents' gender on purchase intention, we created two versions (one for male and one for female) for each of the two conditions (low versus high intensity of craftsmanship). Thus, whatever the randomly assigned scenario was, respondents were in any case exposed to the version of the product compatible with their gender, which has been previously asked and operationalized through a screening question. Both wallets have been selected among the ones displayed on Prada's official Web site and have been slightly modified by substituting the original logo with Kéntro's one (see Fig. 1). Once respondents have been exposed to the text and image stimuli, they were asked which artwork they have seen

---

[1] https://www.bain.com/insights/luxury-goods-worldwide-market-study-fall-winter-2018/.

displayed. This question has been used as an attention check, whose failure implied the respondent's exclusion from the sample. Subsequently, perceived durability and purchase intention of the product advertised have been measured. Furthermore, the extent to which respondents consider themselves expert in the artistic field has also been measured. Lastly, socio-demographics information has been collected.

Most of the constructs have been measured through pretested scales, coming from previous studies and duly translated in Italian. Perceived durability was measured recurring to Stone-Romero et al.'s [75] semantic differential scale ($\alpha = 0.85$), measured on 7 points and composed of two pairs of adjectives (not durable/durable and not reliable/reliable). Respondents reported their willingness to buy by expressing their degree of agreement/disagreement (1 = "strongly disagree" and 7 = "strongly agree") with three distinct statements ($\alpha = 0.95$)—"I would buy this product," "I would consider buying this product" and "The probability that I would consider buying this product is high"—drawn by Dodds et al. [76]. Finally, by using two distinct 1-item 7-point scales (1 = "at all" and 7 = "very much") respondents were asked how confident they felt about art and about the specific art genre displayed.

9.2 Results

First of all, no significant difference between men's ($M = 4.15$ and SD $= 1.51$) and women's ($M = 3.94$ and SD $= 1.70$) purchase intentions has been detected ($t(213) = -0.97$ ns), confirming that differentiating the product advertised by gender has not produced a bias in terms of purchase intention, but has rather prevented it. The mediation model included intensity of artist's craftmanship as the independent variable (high-intensity condition $= 1$ and low-intensity condition $= 0$), purchase intention as the dependent variable and perceived durability as the mediating variable. First, we verified through a regression analysis that the overall effect, that is the positive effect of intensity of artist's craftmanship on purchase intention, was positive and significant ($b = 0.85$, $t(213) = 3.99$, $p < 0.01$). Then, using the bootstraping method as implemented in the PROCESS SPSS Macro by Hayes [77, Model 4], we performed the mediation analysis, which confirmed that perceived durability fully mediates the relationship between intensities of artist's craftmanship and purchase intention. As expected, the effect of intensity of artist's craftmanship on perceived durability is positive and significant ($b = 2.08$, $t(213) = 14.50$, $p < 0.01$), as well as the effect of perceived durability on purchase intention ($b = 0.37$, $t(213) = 3.73$, $p < 0.01$). As expected, the direct effect—that is the effect of intensity of artist's craftmanship on purchase intention controlling perceived durability—is not significant ($b = 0.08$, $t(213) = 0.29$ ns). Thus, both H1 and H2 are supported. Data, however, also shows a significant difference between the two scenarios in terms of respondents' knowledge about the art genre they have been exposed to ($t(213) = 3.68$, $p < 0.01$). Specifically, the level of knowledge of Renaissance art expressed by respondents exposed to the

Renaissance condition (high-intensity condition) ($M = 3.58$, SD $= 1.32$) is significantly higher than the level of knowledge of abstract art expressed by respondents exposed to the abstract condition (low-intensity condition) ($M = 2.90$, SD $= 1.37$).

In sum, the experimental study was conducted to show how, in the context in which a luxury product is advertised through the use of visual art, the level of craftmanship involved in the process of artwork creation can affect consumers' purchase intention and how this effect is mediated by the perceived durability of the product itself. Mediation analysis confirmed these assumptions and demonstrated that perceived durability is responsible for most of the underlying effect. In other words, the study has shown that, when a visual artwork is used to promote a luxury product, the intensity of artist's craftmanship positively affects consumers' purchase intention mainly because it elicits a higher perception of product durability in consumers' mind (Fig. 2).

Low intensity of artist' craftmanship condition   High intensity of artist' craftmanship condition

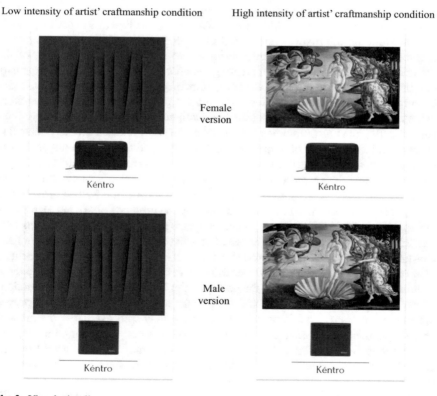

Fig. 2 Visual stimuli

10   General Discussion

The present work pursued two main objectives. On the one hand, it wanted to provide a comprehensive yet concise overview of the luxury phenomenon, looking further into its definition and evolution, as well as into the challenges it currently needs to deal with. On the other hand, it wanted to further investigate the *art infusion effect* under the lens of two luxury-related concepts—namely craftmanship and product durability—in order to enrich the existing theoretical knowledge and provide some relevant insights that could be used by luxury firms to tackle the above-mentioned challenges—namely the ones arisen by luxury "democratization" and by individuals' growing concern toward sustainability.

The *art infusion effect*—namely the positive influence that the presence of art has on consumers' perceptions and evaluations of products with which it is associated—has been investigated by several authoritative scholars, who first proved its existence and then acquired a deeper understanding of it by exploring its underlying mechanisms, as well as its field of action. We joined this line of research by focusing our study on the luxury context and by exploring the effects of an art-related characteristic that has never been investigated so far—that is the intensity of the craftmanship used by the artist in creating the artwork. We found that the intensity of artist's craftmanship is able to affect consumers' perceived durability of the product advertised which, in turn, affects consumers' purchase intention. We therefore believe that the current study can give two main contributions to the existing literature. Firstly, most of the studies conducted so far have focused on the effects that art can produce in terms of consumers' perceptions and evaluations rather than in terms of consumers' behavioral response. As far as we know, only Hüttl-Maack [63] have proved that the presence of art can produce behavioral effects as well, specifically on consumers' willingness to pay. Our study goes one step further, proving that consumers' behavior can be affected not only by the presence of art, but also by the manipulation of the artwork itself. Specifically, we proved that pairing products with artworks that have required a higher level of artist's craftmanship positively affects consumers' intention to buy the product advertised. Furthermore, general agreement exists regarding the fact that the *art infusion effect* occurs because of the spillover of positive properties—specifically, the perception of luxury and the positive emotions arisen by art—from the artwork onto the product advertised. Our study has proved that another property, that is perceived durability, may spill over from the artwork onto the product. In other words, we argue that besides the perception of luxury and the emotions evoked by art, also perceived durability may be transferred from the artwork to the product advertised, leading to positive effects in terms of consumers' response.

The main findings of this research, combined with the knowledge acquired in the process of the literature review, also provide interesting insights for the luxury sector overall as well as practical guidance for marketers who operate in the luxury business environment, which is currently facing important changes and that consequently needs to find new ways to preserve its value in the eyes of its customer base. As proven by our study, leveraging on both art and craftmanship means to elicit perceptions of product durability and thus enhance consumers' purchase intention. In light of the previously mentioned issues affecting the luxury sector, we believe that these findings could be relevant to two main reasons. Firstly, as claimed by Kapferer [25, 38], art is per se able to restore luxury attractiveness in the eyes of the élite and to protect luxury from its democratization. Stressing the craftmanship inherent in the artwork and thus eliciting perceptions of product durability may even enhance this positive effect, since it recalls one of the founding principles of luxury—that is the eternity of its value, which the passing of time cannot affect. Secondly, eliciting perceptions of product durability can also help luxury firms to be perceived as more sustainable and to consequently defuse the moral criticism fueled by the recent overexposure of luxury brands. Indeed, high product durability resounds with the concepts of "slow production" and "slow consumption," two practices that are known to be highly compatible with sustainable development because of their commitment in the protection of both natural and human resources. From a strictly managerial perspective, it follows that it is firstly recommended to continue the ongoing process of artification undertaken by the luxury sector. The incorporation of artistic elements in luxury visual advertising is today more relevant than ever, given the increasing importance that visual communication has gained thanks to the dominant role of social networks. It is secondly recommended to carefully select the type of artwork to display, as well as to privilege the ones in which the intensity of artist's craftmanship is higher, possibly making also use of claims recalling it. Lastly, given all the above-mentioned benefits related to product durability, we also recommend luxury firms to leverage on all the means at their disposal besides art to communicate it (such as packaging and quality certifications).

The present work certainly features some limitations which open the way to further investigations. The most relevant criticality concerns the experiment itself and specifically the stimuli used to manipulate the independent variable. Indeed, since the two levels of intensity of artist's craftmanship (low vs high) have been simulated by resorting to two different types of visual art (abstract vs Renaissance art), findings may be affected by respondents' level of knowledge and liking of the type of visual art they have been exposed to. Indeed, among the two scenarios developed to manipulate the intensity of artist's craftmanship, respondents showed significantly different levels of knowledge of the specific art genre they have been exposed to. In particular, respondents exposed to the Renaissance condition showed a higher level of knowledge of the art displayed than the respondents exposed to the abstract condition. Consequently, we suggest to researchers who desire to further validate our findings to do that by controlling the above-mentioned external variables. Furthermore, the experiment has been conducted by taking into consideration only one product type, while it would be interesting to verify if any changes in the results occur for different types of products. Moreover, considering the strict link between product durability and sustainability, we also suggest to further investigate the effects that product durability may have on consumers' perceptions and behaviors, in order to substantiate the managerial implications we have inferred. Finally, given the consistent variety of tools today used by firms to convey their visual messages, we also recommend to further investigate how our findings may be affected by the use of different means of communication (i.e., billboards, social media, online banners, press, store merchandising).

References

1. Fionda AM, Moore CM (2009) The anatomy of the luxury fashion brand. J Brand Manag 16(5/6):347–363
2. Vickers JS, Renand F (2003) The marketing of luxury goods: an exploratory study—three conceptual dimensions. Mark Rev 3(4):459–478
3. Belk RW (1999) Leaping luxuries and transitional consumers. In: Batra R (ed) Marketing issues in transition economies. Kluwer Academic Press, Boston, MA, pp 41–54
4. McKinsey (1990) The luxury industry. Mc Kinsey, Paris
5. Kapferer JN (1997) Managing luxury brands. J Brand Manag 4(4):251–260
6. Ko E, Costello JP, Taylor CR (2019) What is a luxury brand? A new definition and review of the literature. J Bus Res 99:405–413
7. Mortelmans D (2005) Sign values in process of distinction: the concept of luxury. Semiotica 157(1/4):497–520
8. Amatulli C, Guido G (2011) Determinants of purchasing intention for luxury fashion goods in the Italian market. J Fash Mark Manag 15(1):123–136
9. Amatulli C, Guido G (2012) Externalised versus internalised consumption of luxury goods: propositions and implications for luxury retail marketing. Int Rev Retail, Distrib Consum Res 22(2):189–207
10. Bauer M, Wallpach S, Hemetsberger A (2011) 'My Little Luxury': a consumer-centred, experiential view. ZFP—J Res Manage 38(1):57–67
11. Mauss M (1925) The gift: forms and functions of exchange in archaic societies. Free Press, New York
12. Bliege Bird R, Smith E (2005) Signaling theory, strategic interaction, and symbolic capital. Curr Anthropol 46(2):221–248
13. Veblen T (1899) The theory of the leisure class. McMillan Co., New York
14. O'Cass A, McEwen E (2004) Exploring consumer status and conspicuous consumption. J Consum Behav 4(1):25–39
15. Festinger L (1954) A theory of social comparison processes. Hum Relat 7(2):117–140
16. Snyder CR, Fromkin HL (1977) Abnormality as a positive characteristic: the development and validation of a scale measuring need for uniqueness. J Abnorm Psychol 86:518–527
17. Rosenberg M (1979) Conceiving the Self. Basic Books, New York
18. Belk RW (1988) Possessions and the extended self. J Consum Res 15(2):139–168
19. Eastman JK, Eastman KL (2015) Conceptualizing a model of status consumption theory: an exploration of the antecedents and consequences of the motivation to consume for status. Mark Manag J 25(1):1–16
20. Wiedemann KP, Hennigs N, Siebels A (2009) Value-based segmentation of luxury consumption behavior. Psychol Mark 26(7):625–651
21. Smith JB, Colgate M (2007) Customer value creation: a practical framework. J Mark Theory Pract 15(1):7–23
22. Kastanakis MN, Balabanis G (2012) Between the mass and the class: antecedents of the "bandwagon" luxury consumption behavior. J Bus Res 65(10):1399–1407
23. Vigneron F, Johnson LW (2004) Measuring perceptions of brand luxury. J Brand Manag 11(6):484–506
24. Dubois B, Paternault C (1995) Understanding the world of international luxury brands: the "dream formula". J Advert Res 35(4):69–76

25. Kapferer JN (2012) Abundant rarity: the key to luxury growth. Bus Horiz 55(5):453–462
26. Il Sole 24 Ore (2019). Lamborghini, tetto alle vendite: non più di 8mila auto all'anno dal 2020. (https://www.ilsole24ore.com/art/motori/2019-01-30/lamborghini-tetto-vendite-non-piu-8000-auto-all-anno-2020-103519.shtml?uuid=AFDgn1C&refresh_ce=1)
27. Haute L (2019) Louis vuitton launches ArtyCapucines limited edition handbags in collaboration with renowned global artists. (https://hauteliving.com/2019/06/louis-vuitton-artycapucines/670754/)
28. World Commission on Environment and Development (1987) Our common future. Oxford University Press, Oxford
29. Cooper T (ed) (2010) Longer lasting products: alternatives to the throwaway society. Gower, Farnham
30. Bauman Z (2005) Liquid life. Polity Press, Cambridge
31. Amatulli C, De Angelis M, Costabile M, Guido G (2017) Sustainable luxury brands: evidence from research and implications for managers. Springer, London
32. Carrier JG, Luetchford P (2012) Ethical consumption: social value and economic practice. Berghahn Books, New York
33. De Angelis M, Amatulli C (2018) Luxury marketing: vendere lusso nell'epoca della sostenibilità. Luiss University Press, Roma
34. Hennigs N, Wiedmann KP, Klarmann C, Behrens S (2013) Sustainability as part of the luxury essence: delivering value through social and environmental excellence. J Corp Citizship 52(4):25–35
35. Kapferer JN (2010) All that glitters is not green: the challenge of sustainable luxury. Eur Bus Rev 40–45
36. Bain & Company (2018) Luxury goods worldwide market study, Fall–Winter 2018. https://www.bain.com/contentassets/8df501b9f8d6442eba00040246c6b4f9/bain_digest__luxury_goods_worldwide_market_study_fall_winter_2018.pdf. Accessed 20 May 2019
37. Boston Consulting Group (2019) 2019 true-luxury global consumer insight, 6th edition. http://media-publications.bcg.com/france/True-Luxury%20Global%20Consumer%20Insight%202019%20-%20Plenary%20-%20vMedia.pdf
38. Kapferer JN (2014) The artification of luxury: from artisans to artists. Bus Horiz 57(3):371–380
39. Dubois B, Czellar S, Laurent G (2005) Consumer segments based on attitudes toward luxury: empirical evidence from twenty countries. Mark Lett 16(2):115–128
40. McNeil P, Riello G (2016) Luxury: a rich history. Oxford University Press, Oxford
41. Diniz C, Atwal G, Bryson D (2014) Understanding the Brazilian luxury consumer. Lux Brand Emerg Mark 7–16
42. Yeoman I, McMahon-Beattie U (2014) Exclusivity: the future of luxury. J Revenue Pricing Manag 13(1):12–22
43. Chailan C (2018) Art as a means to recreate luxury brands' rarity and value. J Bus Res 85:414–423
44. Kastner OL (2014) When luxury meets art. Forms of collaboration between luxury brands and the arts. Springer Gabler, Wiesbaden
45. Gibson R (2003) Schiaparelli, Surrealism and the Desk Suit. J Costume Soc Am 30(1):48–58
46. Luxury Society (2009) The evolving marriage of art and fashion. (https://www.luxurysociety.com/en/articles/2009/04/the-evolving-marriage-of-art-and-fashion/)
47. Kim SB (1998) Is fashion art? Fashion theory. J Dress, Body Cult 2(1):51–71
48. Chevalier M, Mazzalovo G (2012) Luxury brand management. A world of privilege 2nd edn. Hoboken, New Jersey
49. Naukkarinen O (2012) variations on artification. contemporary aesthetics, Special Volume 4
50. Joy A, Wang JJ, Chan TS, Sherry JF Jr, Cui G (2014) M (Art) worlds: consumer perceptions of how luxury brand stores become art institutions. J Retail 90(3):347–364
51. Crader S, Zaichkowsky JL (2007) The art of marketing. In: Lowrey TM (ed) Brick & Mortar shopping in the 21st Century, 1st edn. Psychology Press, New York, pp 87–106
52. Hagtvedt H, Patrick VM (2008) Art infusion: the influence of visual art on the perception and evaluation of consumer products. J Mark Res 45(3):379–389

53. Lee HC, Chen WC, Wang CW (2015) The role of visual art in enhancing perceived prestige of luxury brands. Mark Lett 26(4):593–606
54. Ahluwalia R, Rao Unnava H, Burnkrant RE (2001) The moderating role of commitment on the spillover effect of marketing communications. J Mark Res 38(4):458–470
55. Lutz RJ (1975) First-order and second-order cognitive effects in attitude change. Commun Res 2(3):289–299
56. Estes Z, Brotto L, Busacca B (2018) The value of art in marketing: an emotion-based model of how artworks in ads improve product evaluations. J Bus Res 85:396–405
57. Holbrook MB, Batra R (1987) Assessing the role of emotions as mediators of consumer responses to advertising. J Consum Res 14(3):404–420
58. Hirschman EC, Holbrook MB (1982) Hedonic Consumption: emerging Concepts, Methods and Propositions. J Mark 46(3):92–101
59. Strahilevitz MA, Myers JG (1998) Donations to charity as purchase incentives: how well they work may depend on what you are trying to sell. J Consum Res 24(4):434–446
60. Dhar R, Wertenbroch K (2000) Consumer choice between hedonic and utilitarian goods. J Mark Res 37(1):60–71
61. Collins AM, Loftus EF (1975) A spreading activation theory of semantic processing. Psychol Rev 82(6):407–428
62. Huettl V, Gierl H (2012) Visual art in advertising: the effects of utilitarian versus hedonic product positioning and price information. Mark Lett 23(3):893–904
63. Hüttl-Maack V (2018) Visual art in advertising: new insights on the role of consumers' art interest and its interplay with the hedonic value of the advertised product. J Prod Brand Manag 27(3):262–276
64. Simonson I, Drolet A (2004) Anchoring effects on consumers' willingness-to-pay and willingness-to-accept. J Consum Res 31(3):681–690
65. Hagtvedt H, Patrick VM (2008) Art and the brand: the role of visual art in enhancing brand extendibility. J Consum Psychol 18(3):212–222
66. Aaker DA, Keller KL (1990) Consumer evaluations of brand extensions. J Mark 54(1):27–41
67. Bottomley PA, Holden SJS (2001) Do we really know how consumers evaluate brand extensions? Empirical generalizations based on secondary analysis of eight studies. J Mark Res 38(November):494–500
68. Lacey S, Hagtvedt H, Patrick VM, Anderson A, Stilla R, Deshpande G et al (2011) Art for reward's sake: visual art recruits ventral striatum. NeuroImage 55(1):420–433
69. Schultz W (2006) Behavioral theories and the neurophysiology of reward. Annu Rev Psychol 57:87–115
70. Hagtvedt H, Patrick VM (2011) Turning art into mere illustration: concretizing art renders its influence context dependent. Pers Soc Psychol Bull 37(12):1624–1632
71. Peluso AM, Pino G, Amatulli C, Guido G (2017) Luxury advertising and recognizable artworks: new insights on the "art infusion" effect. Eur J Mark 51(11/12):2192–2206
72. Moulard JG, Rice HD, Garrity CP, Mangus SM (2014) Artist authenticity: how artists' passion and commitment shape consumers' perceptions and behavioral intentions across gender. Psychol Mark 31(8):576–590
73. Scarpaci JL, Coupey E, Reed SD (2018) Artists as cultural icons: the icon myth transfer effect as a heuristic for cultural branding. J Prod Brand Manag 27(3):320–333
74. Sheth JN, Newman BI, Gross BL (1991) Why we buy what we buy: a theory of consumption values. J Bus Res 22(2):159–170
75. Stone-Romero EF, Stone DL, Grewal D (1997) Development of a multidimensional measure of perceived product quality. J Qual Manag 2(1):87–111
76. Dodds WB, Monroe KB, Grewal D (1991) Effects of price, brand, and store information on buyers' product evaluations. J Mark Res 28(3):307–319
77. Hayes AF (2017) Introduction to mediation, moderation, and conditional process analysis: a regression-based approach. Guilford Publications, New York

Lasting Luxury: Arts and Crafts with Xia Bu 夏布, a Traditional Handloomed Ramie Fabric

Ying Luo

Abstract Since the dawn of mankind luxury was characterized by the use of expensive materials which skilled craftsmen had carefully reshaped. Since the industrial revolution machines have often substituted craftmanship. This reduction in cost however contradicts the meaning of luxury. Sustainable materials can revert this development: machinery often is not able to replicate skilled craftsmen when processing delicate natural fabrics. We exemplify this trend by showcasing Xia Bu, a traditional hand-made natural fiber from ramie plant that has clothed emperors and commoners for several millennia.

1 Lasting Luxury

In the past decades, when we talked about luxury fashion, we thought of mink coats, leopard bags or luxurious brands full of clothing made of synthetic fibres. As the voices to protect animals and stop plastic pollution are louder and louder, where are those luxuries going?

The picture below shows a noble woman's clothing from Han Dynasty (206 BC–220 AD).[1] It is made of natural fibre and no colour. The thread is very thin and light. The whole garment weighs 49 grams. Its craftsmanship was exquisite and that was made by hands 2000 years ago. "It is light as smoke and thin as a feather". That is how it was introduced in the museum. After 2000 years, it is still luxurious to us. Today people are concerning about the global warming, the sustainability of materials to the environment and the fairness of the modern production to its workers. They want to know where their clothing are made in. I hope this article can remind people to ask what our clothings are made with and how (Photo 1).

---

[1] http://www.sohu.com/a/216387770_100028727 accessed on 18/08/2019.

Y. Luo (✉)
Anthyia Inc., Luxembourg, Luxembourg
e-mail: ying.luo@anthyia.com
URL: https://www.anthyia.com

© Springer Nature Singapore Pte Ltd. 2020
M. Á. Gardetti and I. Coste-Manière (eds.), *Sustainable Luxury and Craftsmanship*,
Environmental Footprints and Eco-design of Products and Processes,
https://doi.org/10.1007/978-981-15-3769-1_5

Photo 1 Unearthed garment hand made in Han dynasty (221–206 BC)

2 Sustainable Luxury Materials

Luxury fashion starts with luxury materials. In ancient China, people treasured the beauty of jade and emerald, worshipped the softness and smoothness of silk and marvelled at the incredible lightness of ramie clothing. With a precious material, a skilful artisan can create art. When we use these luxuries, we treat them carefully and with respect. Our appreciation adds to its value. To start a real luxury fashion, you need these materials. When a young girl is used to fashions made out of polyester, easy to get and easy to discard, she will not understand the beauty of owning a lasting piece. A lasting expensive piece of clothing is not just more comfortable and more expensive, it carries memories and stories. The founder of Apple Steve Jobs said "Some people say, "Give the customers what they want. But that's not my approach. Our job is to figure out what they're going to want before they do".[2] Therefore the power is still in the hands of the designers.

2.1 A Gift from Nature: Ramie

Ramie, called Zhu Ma (苎麻) in Chinese, is an Asian flowering plant of the nettle plant family. It is native to tropical East Asia and is traditionally cultivated in Asia.

---

[2]https://www.goodreads.com/quotes/988332-some-people-say-give-the-customers-what-they-want-but accessed on 18/08/2019.

The ramie fragments have been discovered in the Neolithic sites dating back to 4700 years ago. In the Tang Dynasty, which has been more than 1000 years old, ramie textiles are full of variety and variety, and Xia Bu has been exported outside China, such as Japan and North Korea in much earlier time, and in the eighteenth and nineteenth centuries, it was introduced to many European and American countries.[3] Ramie is also a bast fibre, the plant is grown for its fibres, such as linen, hemp and jute. Bast fibre traditionally is used to make paper, cloth or rope.[4] There are different types of ramie plants. The ramie plant growing along Yangtze River has the highest fibre count. The Yangtze River crosses the provinces of Sichuan, Hubei, Hunan and Jiangxi where most ramie villages are found. Ramie enjoys a tropical climate with high humidity and temperatures (Photo 2).

Ramie has a heart-shaped leaf and it is as big as a whole palm of an adult male man. The matured ramie plant has huge roots. The root can protect the soil along the banks of the great river and are also used in Chinese medicine. Because of its protective qualities for the soil, the Chinese government has made ramie planting mandatory in these regions. Ramie plant is undemanding: it grows very fast, within two to three months it can already be taller than an average human; it does not require irrigation and drainage since it gets sufficient water from the rain. On hills, near ponds and even in the front or back yard it can thrive well without pesticides through the tropical weather. Farmers likes to grow ramie on hilltops where is difficult to grow

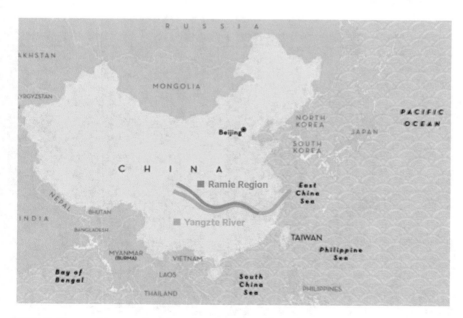

Photo 2 Georgraphic map of ramie region along Changjiang river

---

[3] http://www.cnki.com.cn/Article/CJFDTotal-SJNY199208011.htm accessed on 27/10/2019.
[4] https://en.wikipedia.org/wiki/Fiber_crop accessed on 18/08/2019.

corn or rice. Average ramie crop has 3–4 harvests per year, considering most plants have 2 harvests per year (Photos 3 and 4).

Photo 3 Ramie plant

Photo 4 Ramie field in Da Zhu county, Sichuan province

Photo 5 A man is peeling the skin from ramie stems

In ramie region, the plant creates path for the poor to the middle class: During my meeting with local government officials and bankers, several of them told me the same story: their mothers or their grandmothers supported their higher education with the money made from ramie work at home, such as peeling the stems, drying fibres or collecting wild ramie plants (Photos 5 and 6).

Ramie fibre is one of the earliest fibres that human beings used for their textiles. It is the important material for textile industry for centuries in Southern China. There is an old folk song from *The Book of Songs, also called Shijing*[5] *(which contains folk songs between 800 and 600 BC in Southern China).* From the song, we know that ramie fibre making is a daily activity in those days.

东门之池，可以沤麻。
彼美淑姬，可与晤歌[6]
By the pond at the Eastern gate,
people soak ramie;
With that beautiful girl,
I sing love songs.

At least in olden times, luxury ramie promoted love.

---

[5]3-9-2015 Philology, anthropology and poetry in Arthur Waley's translation of the Shijing Qingyang LIN.
[6]https://so.gushiwen.org/mingju/juv_186bda201291.aspx.

Photo 6 Peeled ramie fibers drying under the sun

3   Ramie Fibre: A Luxury Vegan Silk

Many consumers in the west favourably compare ramie to linen, however, ramie possess a silky sheen unmatched by other bast fibre. This shining lustre results from the length of ramie fibre which surpasses its bast challenges. After years of research and improving industrial clean production of ramie fibre's comfort in wearing; industrial made ramie fibres from Sichuan Province can be as beautiful as silk:delicate, smooth, soft and comfortable, with a silky lustre.

This article wants to present here is ramie's former life: Xia Bu 夏布: the handmade fabric, before the arrival of industrial revolution?

4   Xia Bu 夏布, a Handmade Fabric with Ramie

Xia Bu is hand-woven fabrics with ramie. Xia Bu is coarser than ordinary ramie, because Xia Bu fibre is not being de-gummed.[7] Before the industrial revolution, humans made their textile with hand loomed machines. Xia Bu was one of them with ramie fibres from ramie plants. Although Chang Jiang River region is the centre of ramie cultivation, ramie plants also can be found in the north of the river, from Anhui Province to Korea and Japan, and in the south of the river, from Guangdong Province

---

[7]For ramie dygumming processed fiber, please visit www.ramie.info.

to Vietnam and Malaysia. These ramie plants are different types of ramie plants with less fibre counts. There are many regions with traditional Xia Bu production, most of them are spreading around the main ramie cultivation centre: the Chang Jiang River. The famous ones are WanZhai (万载) region in Jiangxi Province, Liuyang (浏阳) region in Hunan Province and RongChang (荣昌) region in Sichuang Province. Outside China there are also regions that make Xia Bu with different names, such as Echigo Jofu in Uonuma region in Japan, and Mosi in Hansan region in South Korea.

As early as the spring and autumn period and the Warring States Period (770–221 BC), the ancestors of Jiangxi had engaged in the cultivation of ramie and mastered hand-woven fabrics.[8] In Chinese, "Xia" means summer, and "Bu" means cloth. Ramie fibre is cool to the touch, absorbent and breathable. During the Tang and Song Dynasties (618–907 AD), Jiangxi made the finest Xia Bu fabric for ancient Chinese costumes, especially in summer time. It was a gift selected for the emperor as a sign of allegiance. The well-known poet in Tang Dynasty Li bai (also called Li Tai Bo) wrote one poem called Bai Zhu Ci (白苎辞[9]). In the poem, Li Bai described the ramie clothes worn by the female dancers "Light as clouds, with colour as silver" "质如轻云色如银". From this, we could imagine how light those Xia Bu fabrics were. The elements of white ramie clothing appeared more frequently in poems of Song Dynasty (960–1279). In the poem from Dai Fu Gu (a famous poet in Song Dynasty) White Ramie Song (白苎歌)[10]: Dai wrote "…Snow is the weft and jade is the warp, under speeding hands, woven into a piece of ice". 雪为纬, 玉为经。一织三涤手, 织成一片冰", Dai described the beautiful scenery during ramie weaving.

Except using for luxury clothing's at special events for rich people, high-graded Xia Bu was a very popular fabric for wedding and funeral usage in different areas. Long time ago, people discovered that ramie is anti-bacterial, rich people had a corpse wrapped in ramie, hoping to preserve it. Scientists discovered mummies made with ramie in several places in China.[11] In Egypt, ramie was also discovered in Pharaoh's tomb.[12] In some Korean villages, the tradition of burying loved one with ramie still exists. Each year, tons of handmade raw Xia Bu are exported to South Korea. This tradition is also kept in some places in southern China. When a woman marries, she needs to use the red pocket made by Xia Bu. That is a symbol of luck and happiness. In funeral, people wear the mourning dress and caps made with ramie.[13] Wearing ramie in funeral is a tradition to respect your past ancestors.

During the Ming (1368–1644) and Qing (1616–1912) Dynasties, Xia Bu from Jiangxi Provinces were well known in China and abroad. It was exported to North

---

[8]https://baike.baidu.com/item/%E5%A4%8F%E5%B8%83/3477223?fr=aladdin accessed on 18/08/2019.

[9]https://baike.baidu.com/item/%E7%99%BD%E7%BA%BB%E8%BE%9E%E4%B8%89%E9%A6%96 accessed on 18/08/2019.

[10]https://hanyu.baidu.com/shici/detail?pid=283557c5860e3d763366b89d5061cf25&from=kg0 accessed on 18/08/2019.

[11]https://baike.baidu.com/item/%E5%A4%8F%E5%B8%83/3477223 accessed on 18/08/2019.

[12]Seite 149 Ancient Egyptian Materials and Industries.

[13]https://baike.baidu.com/item/%E5%A4%8F%E5%B8%83/3477223 accessed on 18/08/2019.

Photo 7 Xia Bu fiber from Anthyia

Korea and south-east Asia. By the end of the Qing Dynasty, Xia Bu was the main export materials in Sichuan Province. In 1915, on the San Francisco Pacific World Expo, among all the attending countries, China won the most awards for craftsmanship, Xia Bu is one of them.[14] By 1920s and 1930s, Xia Bu cloth industry shrank due to the rise of rayon.[15]

Following the movement of sustainable lifestyle, Xia Bu gained its new life: With easy transportation and communication, different Xia Bu regions are working together closely than ever before. The regions along Chang Jiang River owe fibre rich ramie sources, while Japan and Korea colleagues have good quality sense and beauty eyes for ramie products. Different kinds of workshops, cooperation and trade shows spring up like mushrooms (Photo 7).

5   Ramie Raw Fibre Preparation

After a ramie plant is cut down, the farmer will peeled its skin of the stalk while it is still fresh. Depends on the region, the next step is either drying the peeled fibre in open air or soaking it in water till it becomes soft. (For industry used ramie fibre, farmers only dry them in the air and then sell the dry fibres to factories for de-gumming.)

---

[14] A Sense of Wonder: The 1915 San Francisco World's Fair 7 June 2002–22 September 2002.
[15] https://baike.baidu.com/item/%E5%A4%8F%E5%B8%83/3477223.

6 Xia Bu Yarns Preparation

Each Xia Bu village has its own method of processing ramie, such as bleaching with clear water, daylight rinsing, soaking in due, lime water leaching and charcoal smoke. Regardless of the method used, it takes time. Afterwards, the fibres are sorted by grades: farmers distinguish between the following classifications: Zhuang, first Zhuang, second Zhuang, third Zhuang, Baisuo and Tanqing.[16] According to different grades, fibres are packed differently for the next step.

The next step includes tearing, rolling, crepe and winding. These works are done by female worker. They first torn the fibres into pieces and roll them together; then, put the rolls in a water basin and comb them with their fingers into filament; and then placed the fibres on their thighs and twisted the fibres into a fine ramie yarn. In the end, the yarn will be rolled into a ball (Photo 8).

The fibres are unrolled into fine and long strings, and then brush the strings with rice milk. Rice milk is a thick liquid from cooking rice. When we were young, to make our bedding fresh and cool, we rinsed it with diluted rice milk before drying in the air. Once the milk has dried, the strings are smooth and strong. Sometimes, the rice milk will be mixed with vegetable oil. It will strengthen and smooth the fibre much more. The yarns are spun around on a triangle wooden frame, like the yarn cones used in industrial production, and it is called a "Yangjiao". Now, it is ready to be used in looms (Photo 9).

Photo 8 Xia Bu raw yarn roll from Anthyia

---

[16]https://baike.baidu.com/item/%E5%A4%8F%E5%B8%83/3477223?fr=aladdin.

Photo 9 Rice milk brushing during yarn making for Xia Bu

Weaving is divided into four steps: brushing, mounting on machine, adding rice milk and weaving. Weaving high-quality Xia Bu requires special skill. Lesser craftsmen will break the fibre. Normally men do the warping and brushing work, and women weave.

7   New Luxury: Half Handmade Xia Bu

Most traditional made Xia Bu fabrics are used for window or door curtains and rough bed sheets, just like what they did in Japan and Korea. Since ramie fibres along Chang Jiang River have much higher textile counts, by combining traditional hand making and modern machines fibre spinning, half handmade Xia Bu is born: they are not just soft enough to be used for comforter and clothing, but also they are much more silky in appearance: With new technologies, part of the handwork is replaced with machines. Therefore, we can use de-gummed and much fine ramie yarns to make very fine Xia Bu. The semi-handmade Xia Bu can be very thin and light with half cost. This semi-handmade Xia Bu looks good and has good potential for the sustainable luxury future.

8　Environment Protection and Intangible Cultural Heritage from Chinese Government

A decade ago, the local government of Sichuan Province has mandatory for ramie plant growing to protect its river bank. Now, there are many encouraging policies for the local farmers to grow more ramie, mainly because of economic improvement reasons.

UNESCO inscribed the weaving of ramie in 2011 on the Representative List of the Intangible Cultural Heritage of Humanity.[17]

Here is the decision from UNESCO:

R.1: Transmitted from generation to generation, Mosi[18] cloth weaving is a traditional craft that is rooted in the community and provides its practitioners with a sense of identity and continuity;

R.2: Inscription of Mosi weaving on the Representative List could help to enhance global recognition of the diversity of hand-woven textiles, thus promoting the visibility of the intangible cultural heritage and awareness of its significance.

Korea and Japan kept their traditional Xia Bu production well and were earlier in applying the intangible cultural heritage from UNESCO, such as Korean Hansan village and Japanese Uonuma region.[19]

A few families inherited skilled Xia Bu craftsmen in China survived as Master Tan Zhi Xiang. Now, more and more new young craftsmen are born. They learn not only from their local senior masters but also from international workshops with Korea and Japan. Many of these regions are applying intangible cultural heritage from UNESCO Increased demand from consumers help resuscitate traditional ramie manufacturing techniques.

China, with its long history, has rich and diverse intangible cultural heritage. The Chinese Government attaches great importance to the safeguarding of the intangible cultural heritage and fully supports UNESCO's efforts to safeguard the intangible cultural heritage of humanity.

By today, nearly all the Xia Bu regions inside China are protected under Intangible Cultural Heritage in China.[20]

---

[17]https://ich.unesco.org/en/decisions/6.COM/13.45 Access on 18/08/2019.

[18]Mosi: Ramie in Korean.

[19]https://ich.unesco.org accessed 18/08/2019.

[20]https://baike.baidu.com/item/%E9%9D%9E%E7%89%A9%E8%B4%A8%E6%96%87%E5% 8C%96%E9%81%97%E4%BA%A7/271489?fromtitle=%E9%9D%9E%E9%81%97&fromid= 22079563&fr=aladdin accessed 27/10/2019.

9 Xia Bu Products

Today Xia Bu fabrics are mainly used in high-end home textile. This market trend reflects ramie's special features, which land themselves to summer bedding sheets, pillows, upholsteries, handbags, wallpapers and many more.

10 Hand-Loom Weaving

A hand-loom is a simple machine used for weaving. In wooden vertical-shaft looms, the heddles are fixed in place in the shaft. The warp threads pass alternately through a heddle, and through a space between the heddles (the shed), so that raising the shaft raises half the threads and lowering the shaft lowers the same threads. The threads passing through the spaces between the heddles remain in place. This was great invention in the thirteenth century.[21]

Hand weaved Xia Bu sample By Master Tan Zhi Xiang.

Master Tan is the fifth-generation weaver in Liu Yang Village. From the delicate texture, complex patterns and great density of his works, we can be convinced of the hours Master Tan spent in his world. He is taciturn (Photos 10 and 11).

Photo 10 Master Tan is working

---

[21]https://en.wikipedia.org/wiki/Loom accessed on 18/08/2019.

Photo 11 Master Tan's scarf

11 Embroidery

Embroidery is the embellishment of a fabric, leather, paper or other materials by sewing of designs worked in thread with a needle. Xia Bu embroidery is a hand-crafted art that combines Xia Bu art and embroidery art, blends the natural texture of Xia Bu with simple lines in a limited space.

Each embroidered piece from Master Zhang Xiao Hong brings mountains, rivers and animals to life. The skilful needles, simple natural Xia Bu and strong lasting quality draw luxury into a living space: her embroideries can be framed as an art, or add to shades of lamps, front of bags, or top of pillows and curtains (Photos 12 and 13).

12 Semi-Handmade Xia Bu Bedding Set

In summer, to own a set of fine quality Xia Bu bedding is a luxury. The sheet is cool and fresh, the comfort is light and comfortable. It is natural, you feel healthy and satisfied. That is a feeling of luxury.

Embroidered Xia Bu Bedding Set by Anthyia (Photo 14).

Photo 12 Mast Zhang Xiao
Hong is working

13 Sustainability with Ramie/Xia Bu Projects

It is not only the top grading Xia Bu has charm. Since the Xia Bu is made with hands, the production can be done anywhere in ramie villages, without big investments. A small workshop is easy to set up. The middle-grading and low-grading Xia Bu fabrics can be used in making products by any one who is interested in crafts. Local government supports small business to provide services for the local people. Here is one example.

14 Mama Making Project in Rong Chang Village

"Mama Making" is a charity poverty alleviation project set up by the China Women's Development Foundation to train unemployed women in handicraft skills and provide them income through their own hands. Local governments and companies build platforms for mothers to make and sell traditional Xia Bu handicraft products through different kinds of training activities.

Photo 13   Mast Zhang's embroidered handbag

Photo 14   Enda embroidered Xia Bu bedding set

Photo 15 Rong Chang Mama project

Products include exquisite handmade summer cloth with stitching and embroidery, bags, handmade flowers, etc. One mother told me that she learned new skills and made new friends. She joked that she added the Xia Bu bag nice "zen" touches. I wonder whether it was Xia Bu making gave her "zen" in return. With more and more availability of Xia Bu arts and crafts products "zen", the world will need less and less plastic products.

That is the hope (Photo 15).

Luxury Craftsmanship as an Alternative to Building Social Fabric and Preserving Ancestral Knowledge: *A Look at Colombia*

Alejandra Ospina and Ana López

Abstract Colombia is a multicultural and biodiverse country, recognized for its multiethnicity. This multiethnicity is in part communicated through handicrafts that represent an exaltation to living memory and intercultural dialogue, which evokes tradition and favors innovation as a sustainable and productive alternative. Craftsmanship must be understood as the cultural and material expressions of people with rooted and plural identities. These have been taught through an oral tradition, which has emphasized the union between nature and the human being. In contrast to industrial design and its rational connection to modernity, handicrafts maintain an indivisible line of connection between its creator and its creation, continually reflecting the transformation of culture representation into colors, shapes, designs, and other traditional elements. These elements dignify the indigenous communities while making them visible through craft-making. If we look carefully at the conceptualization of handicrafts and luxury and its diverse characteristics, we understand that the basis of each is the production of exclusive, authentic pieces or services, full of values and stories of its creator. In that sense, a close link is created between the creator and his creation throughout the process and manages to transmit it to the world, generating products with significant symbolic value and durability over time. Sustainability, luxury, and ethics initiate a permanent and unalterable dialogue in the reconfiguration of the fashion system. This dialogue allows addressing in a meaningful manner, the creation of pieces aligned with responsible processes allowing environmental and social balance in a society where luxury cannot exist without sustainability.

Keywords Colombia · Sustainable luxury · Crafts · Craftsmanship · Ancestral knowledge · Indigenous communities · Tradition

A. Ospina (✉)
Budapest, Hungary
e-mail: info@tribeco.co

A. López
Medellin, Colombia
e-mail: waliruu@gmail.com

© Springer Nature Singapore Pte Ltd. 2020
M. Á. Gardetti and I. Coste-Manière (eds.), *Sustainable Luxury and Craftsmanship*,
Environmental Footprints and Eco-design of Products and Processes,
https://doi.org/10.1007/978-981-15-3769-1_6

101

1 Crafts and Indigenous Communities in Colombia

Craftsmanship is the representation of the cosmogony of an indigenous community and its integral relationship with the environment. This form of art continuously develops alongside its fundamental community values. These values, like social norms and respect for the territory, are cultivated through the efficient and responsible use of fibers, plants, fruits, and other elements of this creative process. Craftsmanship must be understood as the cultural and material expressions of people with rooted and plural identities. These have been taught through an oral tradition, which has emphasized the union between nature and the human being.

For indigenous people, handicrafts arise as a process of creating a tangible, material recognition from the community and society at large, being an economic transformation in these communities, an effect of globalization [55]. In its doing, it is recognized as a philosophy of life and political act, telling through symbols the history of the territory, its processes of struggle, resistance, as well as sociocultural and political changes. This social focus, pictured along an evolutionary timeline, is exposed through the fabric, colors, *pintas*[1] or shapes in each piece.

The different forms expressed in each of these handmade creations and identify the community and its territory. It assertively represents its culture, through an allegorical reading that allows the community to preserve and respect its idiosyncrasy and the communities doing. "Crafts are made up of shapes or *pintas* that represent a deep knowledge of nature and a transcendental symbolic and cultural content, which leaves a community footprint in history" [7, p. 4].

According to Serrano [54], ancestral knowledge has been consolidated into material pieces to express the diversity of the world and the coexistence of different worldviews as well as the ability of people to organize themselves, while optimizing natural resources. It also expresses the community capacity to build productive processes and social structures that can withstand time and create institutions that allow the communities to establish themselves as a society and generate bidirectional relationships with the State.

This social fabric dynamic understands that craftsmanship is a vehicle to materialize cultural and social expressions, build a collective identity, and energize local economies [51]. These dynamics are interconnected globally through cultural evolution and the transformation of sociopolitical narratives, such that there is a creative symbiosis between the subject and the object.

Each handmade piece is thus a construction and manifestation of community values as well as cultural identification. An art form is capable of weaving to the rhythm of history and withstanding time, thus maintaining its connection with the environment; art forms achieve to maintain its production ways from before the industrial revolution until today (see Fig. 1).

---

[1]It is the word used by indigenous communities to refer to the forms that appear in the crafts and represent geometric figures, animals, plants, and other things.

Fig. 1 Wayúu woman
weaving a backpack or in
their language "Mochila
Wayuu" (Image sent by
María del Pilar Rodríguez,
researcher and creator of the
Awana Taller brand)

For indigenous communities, handicrafts became the ideal way to safeguard ancestral knowledge and to easily link themselves to nature through the generations, slowly and by natural cycles. This art form is composed of a whole set of tangible symbols that express aesthetic vestiges and an artistic richness of a flourishing culture in the remote past and its permanent evolution. Each handicraft manages to keep tradition and ancestry united on an indivisible line. This is the basis of what each handicraft represents for each community and each space that it pertains to [1].

Artisans can look at the environment and understand its production cycles, the evolution of nature, and its ability to regenerate to stay in time. This ability enables artisans to initiate craft creation processes based upon what the environment provides the community.

Indigenous peoples are configured as the only creators that are able to achieve a "rapport and integration with the habitat, to meet their needs, within logical, rational, respectful and harmonious responses and processes" [47, p. 98] while at the same time creating exclusive and connected pieces that are symbolic of and demonstrate an interaction between the craftsman, the territory, and the consumer.

The crafts, as a conceptualization of a people's being, allow the construction of cultural identities, spaces of legitimation, and reproduction of tangible symbols

[14]. Implicitly, they configure a visual and aesthetic narrative, which accounts for the territorial and cultural transformations that have occurred throughout history and constitute a part of the nation's tangible and intangible heritage. This resignifies popular cultures and dignifies collective memory through tradition and handwork.

2 Colombia

Colombia is a multicultural and biodiverse country, recognized for its multiethnicity. This multiethnicity is in part communicated through handicrafts that represent an exaltation to living memory and intercultural dialogue, which evokes tradition and favors innovation as a sustainable and productive alternative.

According to the National Indigenous Organization of Colombia (ONIC), there are 102 indigenous peoples in the country recognized by the Colombian State through the Ethnic Directorate of the Ministry of Interior and Justice, the National Planning Department and the National Statistics Department [45]. These peoples are mainly located in 25 of the 32 departments that make up the national territory[2] (see Fig. 2).

In the country, the craftsmanship sector links more than 350,000 people who work in handicrafts. They represent 15% of manufacturing employment. Additionally, 58% of the population that works in handicrafts is dedicated to weaving and basketry, 13% to woodworking, 8% to jewelry, 7% to ceramics and pottery, 5% to leatherwork, and the remaining 9% represent other trades.[3]

More than 114 species of plants[4] are used to extract natural fiber and carry out basketry, weaving, textiles, and natural dyeing work. In Colombia, there are currently more than ten lines of artistic development that express textual, creative, and physical manufacturing of a new intellectual space. This space contains ancestral wealth that is embodied in multiple ways of doing ancestral crafts [9].

The context of Colombian territorial violence, forced displacement, extreme poverty in some departments of the country and the power play dictated by the dominant institutions, has created conditions of vulnerability, mainly for the Indigenous, Afro-Colombian, Palenqueras, Roma, and Raizales communities.[5] These communities are the part of the immaterial wealth of the nation [43]. From colonization to modernity, they have merged their knowledge and their intercultural development as a source of creation of new platforms for disruptive and sustainable economies [48].

---

[2] Accessed 25 June 2019 https://www.onic.org.co/noticias/2-sin-categoria/1038-pueblos.

[3] Accessed 25 June 2019 https://www.semana.com/especiales-comerciales/articulo/turismo-artesanal-una-mirada-al-patrimonio-cultural-colombiano/429530-3.

[4] Accessed 22 June 2019 http://www.artesaniasdecolombia.com.co/PortalAC/C_noticias/fibras-vegetales-elemento-basico-de-las-artesanias_5079.

[5] Afro-Colombians: people of African descent born in Colombia. Palenqueros: descendants of slaves. Raizal: native population of San Andrés, Providencia and Santa Catalina, descendants of the union between Europeans and African slaves. Roma: ethnic minority group, descended from northern India, recognized in Colombia, through decree 2957, of August 6, 2010, as part of the ethnic and cultural diversity of Colombia.

MAGDALENA
• 1 Nararakajmanta
Ette Ennaka (Chimilas)
SUCRE
• 2 El Higuerón
Afro Descendiente
ANTIOQUIA
• 3 Caimán Alto
Guna Dulce (Cuna)
• 4 Caimán Bajo
Guna Dulce (Cuna)
• 5 Polines
Embera Katío
• 6 Jaikerazabi
Embera Katío
CALDAS
• 7 San Lorenzo
Embera Chamí
CHOCO
• 8 Papayo
Etnia Wounaan.
TOLIMA
• 9 Amayarco
Coyaima (Pijao)
• 10 Coyarcó
Coyaima (Pijao)

CAUCA
• 11 Coop-mujeres
Alto Descendiente
• 12 Canaan
Eperaara Siapidaara
• 13 La María
Misak (Guambiano)
NARIÑO
• 14 Barbacoas.
Afro descendiente
• 15 Resguardo indígena
Gran Sabalo. Awá
PUTUMAYO
• 16 Buena Vista
Siona
• 17 Santa Rosa de Guamuez
Cofán
• 18 Yarinal
Cofán

• 19 Juan Cristóbal
Kichwa
• 20 San Marcelino
Kichwa
• 21 Nuevo Amanecer
Siona
• 22 La Cristalina
Embera Chamí
• 23 Condagua
Inga

GUAJIRA
Maestras artesanas • 24
Wayúu
Kasiwolin y Arrutkajui • 25
Wayúu
Ranchería el Mojan • 26
Wayúu
Ranchería Iwouyáa • 27
Wayúu

CESAR • 28
Resguardo Kankuamo
(Atánquez, Guatapurí, Río
Seco, La Mina, Los Háticos,
Chemesquemena)

NORTE DE SANTANDER
Motilón Barí • 29
Barí
ARAUCA
Playas de Bojavá • 30
U'wa
VICHADA
Muco Mayuraguaí • 31
Sikuani
Cumariana • 32
Sikuani

GUAINÍA
Sabanita • 33
Curripacoi
Cocoviejo • 34
Curripaco
GUAVIARE
Miraflores • 35
Tukano
VAUPÉS
Puerto Tolima • 36
Tukano, Siriano, Cubeo
Villa María • 37
Tukano, Siriano, Cubeo
Puerto Golondrina • 38
Cubeo
AMAZONAS
Puerto Guayabo • 39
Yucuna, Tanimuca, Letuama y Matapí
Nazareth • 40
Tikuna
Macedonia • 41
Tikuna y Cocama.

Fig. 2 Distribution of indigenous communities in the Colombian territory. Image made by the authors of this chapter. Information taken from the research Origins, carried out by Ecopetrol and *Artesanias de Colombia*, 2004

For years, indigenous peoples have been immersed in the Colombian conflict. They have been part of the processes of collective invisibility, violation of rights, uprooting of the land, loss of their customs due to the need to immerse themselves in the city, cultural appropriation by large brands, in addition to the non-recognition of their identity and culture by part of the Colombian territory [3].

The continuous sociopolitical changes, mediated by the large-scale conflict, configure dynamics of physical and intellectual violence centered on vulnerable communities, rural areas, and places of population settlement [38]. Consequently, affecting primarily indigenous peoples through the exploitation and destruction of their land, fauna, and flora, which is their primary center of spiritual connection, creation, and economic and cultural development for them. Here, handicraft acquires a new narrative and exponential nuance, which envisions artisanal and territorial resilience as a space for local production and support for creative actions in dispute spaces [8].

Within this framework of exposure and cultural risk, the *Artesanias de Colombia*[6] program emerged in 1964, and, in 1968, it was declared a mixed economy company linked to the Ministry of Economic Development. It has thus become a promoter of the development of the artisanal sector through programs and spaces of visibilization throughout the Colombian territory. In this way, this program responds to the latent need to protect handicrafts as social construction and representation of the cultural heritage of each region, its expression, communication, and commerce [6].

Handicrafts imply recognition and respect for the local characteristics of the communities and its traditional products, which express and keep alive the culture of each region of the world [29]. This contributes to the identitarian wealth of the nation and the building of a national history through collective memory.

Faced with the need for change and restructuring of the status quo, the Law 36 was enacted in 1984, which recognizes traditional folk crafts as "the production of handicrafts, resulting from the fusion of American, African and European cultures, *produced by the people in anonymous forms*, with complete predominance of the material and the elements of the region, transmitted from generation to generation."[7]

However, the conceptualization of this legislative declaration based on the former Colombian constitution allows communities and the ancestral knowledge that is embodied in their crafts to continue in processes of non-recognition by the Colombian people and to stay subject to the permanent manipulation and violation of their rights as subjects and creative persons, subsequently leaving their work open to cultural appropriation at national and international level.

Six years later, there was an organizational change, envisioned in the 1991 constitution that was articulated in the Culture Law which aimed to recognize the territory as a multiethnic nation, governed by laws that visibility and protect indigenous communities and living traditions, as a constituent part of a homogenous country with a broad cultural diversity [36].

---

[6] www.artesaniasdecolombia.com.

[7] Law 36 of 1984. November 19 of 1984, Law of the Craftsman, Decree 258 of 1987. Chapter 1, article 6.

The wealth of our nation is manifested in ancestral knowledge, in the traditions present in the trades, the use of materials and symbolic expressions that are put at stake in the artisanal exercise [46, p. 6].

- The "Vueltiao" hat of the Zenú community is the Colombian handcrafted piece par excellence at the national and international level, where the transformation of the cañaflecha fiber into totemic figure narrates the community's relationship with the universe and nature for more than a thousand years (Fig. 3).
- The weaving of the *Waleker* (spider) of the Wayúu women is part of the initiation rites of adolescent girls to adult life; the *Kanás* (designs) reveals the matriarchal structure of their society and has become a cultural manifestation (Fig. 4).
- The women of the Arhuaca community make the Arhuaca backpack, which is identified as a symbol of an extension of the body, and the uterus (Fig. 5). Through their knitting, the women of the Arhuaca community express their ancestral knowledge. The backpack is woven in the form of a spiral, the symbol of the creation of the world [50].

Each handmade piece contains intrinsically cultural forms developed over time. These forms are often subject to external changes that permeate and transform the artisan relationship between the artisan and the outside world, thus resignifying the

Fig. 3 "Vueltiao" hat, making by the Zenu community with natural fiber: *cañaflecha*. María del Pilar Rodríguez's picture, worker and connoisseur of crafts and Colombian indigenous communities (Image sent by María del Pilar Rodríguez, researcher and creator of the Awana Taller brand)

Fig. 4 Backpack made by
the women of the Wayuu
community. María del Pilar
Rodríguez's picture, worker
and connoisseur of crafts and
Colombian indigenous
communities (Image sent by
María del Pilar Rodríguez,
researcher and creator of the
Awana Taller brand)

Fig. 5 Backpack made by
the women of the Arhuaca
community. María del Pilar
Rodríguez's picture, worker
and connoisseur of crafts and
Colombian indigenous
communities (Image sent by
María del Pilar Rodríguez,
researcher and creator of the
Awana Taller brand)

sociocultural context of indigenous peoples and the disposition of their knowledge
as a new form of sustainable development.

3 Can Ancestral Crafts Flourish in a Capitalist World?

The first interactions between indigenous people from America with western culture were not friendly nor voluntarily accepted. From the very beginning of this interaction, colonialism forced radical life shifts on peoples. Those shifts involved different social and political models, religions, and even language. Notwithstanding, ancestral communities found means to safeguard their history and traditions. They did it through art, music, and crafts. They embodied their culture in each piece of craft they conceived, and in each piece of clothing, they made and wore. They passed down history to new generations through oral traditions such as craft-making [12].

Currently, the ancestral techniques put into the handmade work of each craft have been subject to permanent rights violations, cultural appropriation, and plagiarism. These violations are mainly due to the general ignorance of ancestral meanings united to a lack of legal protection for its material production. Although their crafts helped them to keep their culture alive, today, these communities fight for respect and protection.

When capitalism is imposed as the only economic system that appears to work, often its strategy is to urge the transformation of community structures. This transformation usually urges new processes that instead of promoting development and independence, promote economic dependence and a culture of humanitarian assistance. These new processes undoubtedly are damaging communities.

Developing countries receiving high quantities of second-hand clothing as part of humanitarian aid programs are an example of dependence promotion programs. These countries rarely receive investments to improve infrastructure, implement environmental policies, promote exports, support local businesses to push local development, or to provide a significant increase in job creation [41].

The promotion of a dependant economy enables, among others, cases of cultural appropriation. In recent years, many have occurred in different Latin American countries, physically and intellectually transgressing indigenous communities and their processes of creation and social transformation.

In the concrete case of the fashion industry, a positive social, economic, and political impact could be generated by paying fair wages and promoting artisanship by acknowledging their authority and ownership over their crafts and ancestral meanings. Nonetheless, their work has been lagged, only recognizing famous designers and big brands, thus forgetting that each element used in a garment makes an essential part of the identity of the maker as well as the wearer [16].

These capitalist dynamics and the lack of visibility and recognition of ancestral practices through crafts have allowed the cultural appropriation of ancestral knowledge by local and foreign actors. Such crafts establish themselves in a market that, although values these art forms, is unaware of its history, spiritual meaning, and creation process. The hands of the person who performs it, and the sense of belonging that it grants to each indigenous community goes unknown [42].

Over recent years, we have seen continual growth in the interest of purchasers to acquire pieces that reflect the history and tradition of ancestral peoples. With this trend, we require a redefinition of rules for the commercialization of these pieces, considering the respect, protection, and promotion of ancestral knowledge. Regrettably, these trends come along with corporations and people taking advantage of certain people's crafts, from people buying at ridiculous prices (taking advantage of the need of some communities) to people just copying designs and techniques and calling it "inspiration" without giving the proper credit to the artist and communities. The appropriate term for these practices is what we know today as cultural appropriation [49].

Cultural appropriation is a term being born recently. The Oxford dictionary included it just in 2017 and defined it as "a term used to describe the taking over of creative or artistic forms, themes, or practices by one cultural group from another." Brigitte Vézina, in her 2019 CIGI Paper "Curbing Cultural Appropriation in the Fashion Industry" defines Cultural Appropriation as "The act by a member of a dominant culture of taking a TCE whose holders belong to a minority culture and repurposing it in a different context, without the authorization, acknowledgement and/or compensation of the TCE holder(s)" [56, p. 6].

4   Sombrero Vueltiao, an Example of Cultural Appropriation

The "Vueltiao" hat has been a flagship piece of the Zenú community and the Colombian people, for more than three thousand years. It is a cultural symbol of the nation and economic base of this community, including the artisans of Tuchin, who learn the technique from an early age.

The use of cañaflecha as the raw material and natural pigments to achieve the dark hue and the process of sowing, harvesting, drying, design, and braiding make of this piece multiple sources of heritage wealth and recognition both locally and nationally.

In 2013, the streets of Cartagena, Colombia, were filled with "Vueltiao" hats made in china. The low price, the use of synthetic materials and artificial dyes, as well as their production origin, alarmed the entire country. China had been responsible for reproducing a cultural object serially stripping it of all its history, its ancestral knowledge, and exclusivity.

The locals mentioned a visit by Chinese businessmen, who have come to know the creation process of the hats. People quickly started selling Chinese hats that looked like real ones but with substandard quality and lower prices. You could buy one of these hats for 5 dollars, while the original price of the Vueltiao hat is above 18 dollars.

That same year, the government signed commitments with the community to help them protect the intellectual property of the hat, which by then had the designation of origin, but had no customs protection or awareness-raising programs in place.

These commitments included awareness programs targeting consumers and sanctions to those who produced and sold the replica of the hats. These awareness programs were linked to the media to achieve a massive impact on locals and tourists. Additionally, a visual campaign was initiated by delivering flyers in the streets of Cartagena to promote interest to know the traceability of the hat, its artisanal process and the cultural richness that each piece intrinsically carries from the fabric.

When it comes to protecting the intellectual and cultural property of handicrafts, government programs that legally protect their ancestral knowledge are essential. However, these are not enough if they are not accompanied by education programs focused on the consumer and a commitment from the private sector that promotes ethical marketing practices.

5 Globalization

The strong demographic growth of the last decades and the current Colombian territorial context connect the handcrafting of indigenous peoples to a framework of urban development, with new forms of cultural, social, and economic relations. These new forms propose structures of urban alienation, articulating them with new forms of capitalist creation and development, thus facilitating the adaptation of ancestral knowledge and material expressions of popular cultures in different axes of large-scale production and commercialization [4]. As a result, craftsmanship is visibly altered in terms of exclusivity, sustainability, and preservation of the environment.

Commercial transculturalization, typical of globalization, and the opening of borders for free trade between countries cause new economic ideas and games of power and dominance. They are consequently shaping forms of production and consumption in times and spaces different from those conceived. Those market displays are interspersed by some indigenous communities, which begin to urbanize and blur the characterization of their work and collective tradition, in the vertiginous rhythms of the city and accelerated production.

Achieving the keeping of the community base as the only form of unification across the territory and handicrafts is only possible by returning to the origin of the villages, and the exaltation of oral tradition plus artisanal manifestations, as a material expression of validity and distinction. This thought of resistance against a society of voracious consumption and production and the acculturation process is the way of resignifying artisanal luxury as a means of exaltation and protection of the tradition of autonomous peoples, forgers of culture [10].

The craftsmanship is presented as a pre-industrial mode of production alternative to modern production, managing to maintain the craft ability to remain in time under its creation and production times, away from the voracious rhythm of modernity [13].

In contrast to industrial design and its rational connection to modernity, handicrafts maintain an indivisible line of connection between its creator and its creation, continually reflecting the transformation of culture representation into colors, shapes, designs, and other traditional elements. These elements dignify the indigenous communities, while making them visible through craft-making. Here, the symbolism of values is updated to maintain collective memory and interconnect with other ways of being and staying in the world [17]. Handicrafts become a luxury experience before, during, and after, symbolically linking the creator, to his environment, and his buyer.

The coexistence and adaptability of handicrafts to the environment manage to maintain the slowness in its processes, the search for natural materials, and the exclusivity in each piece [2]. In the words of Honoré [35], minimizing and slowing down the modern industry, communities can maintain a high-quality production process and generate balance in the development of crafts.

The development of globalization and its structural changes permeate the forms of socialization and creation of indigenous communities, allowing them to rebuild themselves internally and delve into the concept of luxury. Subsequently, trough luxe is possible to safeguard their ancestral knowledge, using the crafts and art forms to make up the wealth of the communities [5].

In this new look, crafts rise to be considered as models of sustainable living that balance the relationship between human beings and nature [52] and connect in a new line, fusing aesthetics and ethics as the new definition of luxury [28].

With modernity and the development of handicrafts from an inside-out look, hands become the primary tool of the productive process where craft-making interconnects with nature and its evolutionary cycles. Present and past emotions, at the individual and collective level, are transformed through symbols, to build local and global dialectics. Even more, initiating an endearing process where exclusive design, aesthetics, respect for the environment and times, make up a luxury craft, based on principles of differentiation and identity, promulgators of a tremendous ancestral and national wealth.

The ancestral wealth, given through symbols in each handmade piece, allows it to be understood as an element of resistance that transforms imaginary and restructures capitalist notions from disruptive economies based on the innovation of materials, permanence, and the relationship with the environment, tradition, and living memory [18]. All of the above added to exclusivity, handwork, and a symbolic relationship between craftsman–consumer represent luxury by itself.

6 Sustainable Luxury as a Vehicle to Savage Ancestral Crafts

Luxe is an ambiguous concept, with different meanings, depending on the perspective and cultural setting used at the time of the definition. Traditionally, "luxury may be defined as an inessential, desirable item that is expensive or difficult to obtain" [33, p. 4].

The definition above may only alienate the concept, eliminating the possibility of reconciliation within the sustainability movement. However, there is a considerable need for regular awareness-raising, while emphasizing the shared values between luxury and sustainability, such as creativity, high-quality, and production of exclusive and timeless pieces, among others.

The luxury fashion industry as a robust value creator and trendsetter can generate a more significant impact on the way we consume nowadays [33]. It can also support the way we see artisanship and craftsmanship. In a society where massively manufactured cheap clothing is the norm, unique pieces, with meanings and stories, are the way to go.

Hence, sustainable luxury opens endless possibilities to support and empower indigenous communities with ancestral knowledge by giving a new meaning to luxury, where the expression of values and beliefs, united to high-quality and handmade work, can play a fundamental role to fighting the copying of crafting skills and designs [24].

Luxury goods have implicit in their construction sustainability characteristics, such as the lasting duration of the piece and the no need to be part of a trend. Luxe goods do not just go out of style. They remain as a vestige of the consumer personality and values.

Even more important, there is a need to return to the essence of luxury and use the connection with craft and its possibilities of recovering ancestral techniques, as well as to promote the production of artisan items, handcrafted in a responsible way and with a vast ancestral meaning behind it [15].

Although luxury has always been important as a social determinant, it is currently starting to allow people to express their intrinsic values. Thus, sustainable luxury promotes a return to the essence of luxury itself with its ancestral meaning, i.e., a thoughtful purchase, artisan manufacturing, beauty of materials in its broadest sense, and the respect for social and environmental issues. Consequently, sustainable luxury would be not only the vehicle for greater respect for the environment and social development, but also a synonym for culture, art, and innovation of different nationalities, maintaining the legacy of local craftsmanship [19].

If we look carefully at the conceptualization of handicrafts and luxury and its diverse characteristics, we understand that the basis of each is the production of exclusive, authentic pieces or services, and full of values and stories of its creator. In that sense, a close link is created between the creator and his creation throughout the process and manages to transmit it to the world, generating products with significant symbolic value and durability over time.

Luxury can offer a unique opportunity to create sustainable business environments due to its two main characteristics that differentiate it from other market segments or industries [27]. First, luxury is (often) based on unique skills; it allows luxury to provide high-quality and rewarding business conditions. Second, luxury is characterized by its unique relationship with time, since its value is everlasting, enabling luxury to offer a sustainable business model for resource management and the development of high-quality products [31].

The fusion of crafts and design, empirical creation and theory, ancestral knowledge, and academy allow cultural immersion to develop proposals that harmoniously blend aesthetics, indigenous elements, the trade of hands and design to create products with high intellectual, emotional, environmental, social, and economic values.

Each of these characteristics mentioned above positions luxury craftsmanship as a product capable of connecting sustainability, beauty, and quality in each stage of its production process. Luxury craftsmanship goes beyond the craft-making, putting first its essential relationship with luxury and sustainability and their ability to maintain culture through art and innovation of different nationalities without homogenizing the legacy of local crafts [19].

To talk about luxury, from a perspective that revalues socially developed concepts, is to talk about crafts and all their creative process from ancestral knowledge as a living tradition of the people to the finished product and all the intangible and natural wealth that it contains in itself.

The cultural wealth associated with luxury is only possible to the extent that each piece can be identified even in a homogenized and mass market by an industry whose ability is to overproduce at a dizzying and exponential rate. The disarticulation of the traditional values of the system proposes an inside look toward individual and collective values, which rescue the identity and origin of the being through experience and recognition of its surroundings, as the main sign of acceptance, uniqueness, and permanence in time [52].

A reconceptualization of the system allows us to understand craftsmanship as cultural luxury, associated with the essentiality of each piece, its emotional and symbolic value, and its creation process. Craftsmanship must base its production on the preservation of heritage and tradition, as a source of transfer of historical and collective memory. This precious transfer is invaluable; human-based "cultural fortunes" are obtained in the process and can be defined as luxury items [32].

The articulation between artisans, craftsmanship, and luxury, from an indivisible triad of artisanal luxury, allows connecting symbolically with the product through the knowledge of where and who made it. The traceability of the craft must be explored as an experience, understanding the craft as a fundamental pillar of the life, culture, and the development of indigenous communities.

The creation of a handmade piece goes beyond carefully chosen materials to maintain the capacity of natural regeneration and the proper balance of natural ecosystems. Artisans have an admirable ability to look inside and recognize the territory as the primary source of raw material, to capture their worldview through manual skills and creative development. The craft-making process is only doable thanks to the communities' ancestral tradition and knowledge.

According to Gardetti [28], the drastic decline of natural resources, the extinction of species, the disappearance of millenary cultures, industrial overproduction, and other devastating effects of the twenty-first century make humankind rethink development. Purchase then lean on luxury products or services created under parameters of social and environmental responsibility. "When we talk about sustainable luxury, we refer to brands that have as a beacon not only create brands of excellence and higher standards but also protect the environment and the communities in which they operate".[8]

The use of natural fibers, such as the *Cumare*, the *cañaflecha*, the *iraca,* the *werregue*, the *paja tetera*, the *chocolatillo*,[9] and other native plants of different Colombian territories, provides the design with an intrinsic connection with a local context that goes beyond its borders and is consolidated as the source of multicultural connection between nations. Fruits, such as achiote, turmeric, vegetables, and plants, have been used anciently by indigenous communities to achieve unique pigments and natural color palettes that connect harmoniously with the created piece and its intellectual value.

Reassessing concepts established in the traditional fashion system, including in making fashion a conceptual look that integrates the creative ability that exists in the country, its diversity and the capacity for the responsible use of natural resources as a source of sustainable raw material, has allowed creating spaces for articulation of ancestral knowledge.

Design and crafts meet permanently to exhibit through exclusive pieces, the history behind a territory, the resilience of a community, artisanal talent, the beauty of Colombian flora, and the ability to maintain dynamic equilibrium and long-term stability systems at the environmental and social level. These microsocial dynamics converge for the recognition of craft-making and the meaning of its creation at the national and international level, maintaining the legacy and distinction of local crafts. These characteristics of diverse societies in race and culture such as Colombia add to the luxury crafts a characteristic of differentiation and uniqueness.

The brand Religare in union with the Coreguaje indigenous community, as well as the designer Álvaro Leyton along with Doris Jajoy, weaver teacher of the Inga community, are two national perspectives addressed in this chapter, to approach luxury crafts as a revolutionary process in the contemporary world. Through these examples is possible to observe individual and collective choices linked to history, culture, the environment, and other forms of organization in the world, achieving a feeling of innovation, balance, and ethics [23].

7 El Cumare

The Coreguaje and Tikuna communities located in the department of Caquetá, south of the Colombian territory, inhabit lands where the chambira, a local palm, grows. This palm has for years represented the basis of its economy and sociocultural development process through the fabric: the Cumare.

This palm is used to make backpacks, *chinchorros*,[10] hammocks, hats, and necklaces of striking colors, achieved with ancestral and natural dyeing techniques, using

---

[8]https://www.essentiaconsulting.net/lujo-sustentable/. Accessed 2 July 2019.

[9]Typical natural fibers of Colombia.

[10]Large hammocks created with a handmade fabric.

Fig. 6 Leonidas Gutiérrez, leader of the Coreguaje community of Caquetá. Luz Marina, community Cumare fiber weaver (Image sent by the designer Manuela Peña, director and founder of the Religare brand. Photographer: Felipe Cuartas)

tree bark, roots, leaves, flowers and fruits of the region. Each part is used for some purpose, ranging from food, medicinal use, as well as initiation ceremonies performed within the community.

The process of planting, collecting, and preparing the fiber is carried out manually, as well as the extraction of threads, dyeing and then the fabric, which is achieved with cross technique, knots or in fabric.[11]

The in-depth territorial knowledge, developed by the communities in each of their environments, has allowed them to establish bonds of belonging and root with the environment around them.

They manage to capture in their products not only their cosmogony but the accumulated knowledge over the years. Indigenous people are intellectual and experienced teachers, who embody sustainable and prosperous ways of life and circular economies, able to reinvent themselves over time to maintain social balance (Figs. 6 and 7).

The Cumare or chambira is one of the most desirable fibers in Colombia, Brazil, Peru, and Ecuador thanks to its strength, flexibility, and durability, and the material and cultural value it has for the present and future generations.

Manuela Peña is the creator of Religare, which means "to unite again." This brand from Antioquia, Colombia, was born with the purpose of returning to nature a bit of what it gives us through design cycles.

Religare has established itself as a sustainable luxury brand, thanks to the articulation with the Coreguajes community and the work of craftsmanship that they do with the Cumare fiber. The work of the brand with the community includes processes of cultural immersion, recognition of the territory, time and spaces of the communities, as well as the co-creation, weaving, and inclusion of conventional pattern techniques of pieces from ancestral knowledge.

---

[11] Accessed 5 July 2019 http://www.waliruu.com/religare/.

Fig. 7 Backpacks made of Cumare fiber by the Coreguaje community, located in the department of Caquetá, Colombia (Image sent by the designer Manuela Peña, director and founder of the Religare brand)

The first step of Religare was knowing the nature of the Cumare fiber within the Colombian fauna and understands the nobility and duration of the material and its sustainable harvest process. The understanding of the fiber would allow the brand to avoid overexploitation of the soil and the extinction of the plant while savaging the significance that the Cumare has for the Coreguaje community. The brand makes sure to preserve their cultivation and establish responsible consumption practices that allow the community to continue using this vegetable fiber.

"The cumare for us is a legacy left to us by our ancestors, as the main raw material for our native fabrics. Our craftsmanship is made with that fiber". Leonidas Gutiérrez, leader of the Coreguaje community of Caquetá.[12]

Proposing, discussing and experiencing other forms of use, was an activity that took time, but it was consolidated through two-way teaching and learning, where Coreguajes made available, fiber management and the art of knitting, and Religare, different textile patterns created. Then the fiber will be molded under the firmness and rigor of the Cumare and its natural beauty (Fig. 8).

Craftsmanship as diversity is enormous and arises from different realities. Therefore, it must be seen and understood as a plurality that is narrated in a particular context through art [54].

---

[12]Personal conversation between Manuela Peña and Leonidas Gutiérrez, shared on August 3, 2019.

Fig. 8 Collar Ñuka, made in Cumare fiber by the Coreguaje community and the Religare brand (Image sent by the designer Manuela Peña, director and founder of the Religare brand. Photographer: Felipe Cuartas)

Before the creative process started, the Coreguajes shared their culture, political organization, connection with the territory and the development of the fabric to develop the backpacks, which are their main craft available to the modern world. This intercultural dialogue allowed the creation of related and collaborative spaces, which started eco-design processes, designed to weave versatile garments, without generating waste or decontextualizing the proper meaning of the raw material.

The creative synergy established between Religare and the Coreguaje community creates a new symmetrical line where aesthetics, ethics, and exclusivity converge. These characteristics typical of sustainable luxe are creating a new space for national visibility and recognition of the cultural wealth and immaterial legacy inscribed in craftsmanship as well as the Cumare's participation in the fashion industry (Fig. 9).

The craftsmanship, explored under the concept of luxury, sophistication, and high quality of materials and pieces, allows the creators to find new forms of exhibition, responsibly exposing ancestral knowledge in the design and fashion industry.

The co-creation processes, carried out between Religare and the Coreguaje community, manage to integrate the knowledge and craftsmanship of indigenous peoples

Fig. 9 Top in Cumare
(Image sent by the designer
Manuela Peña, director and
founder of the Religare
brand)

in one piece, through the recognition of each garment as the materialization of the know-how of a particular community. Religare manages to protect the ancestral knowledge and environmental capacity of the community through the recognition, respect, and development of work ties. By sustaining its work on the intellectual and economic reciprocity with the communities, it is possible to make them visible, dignify them, and recognize their ancestral knowledge.

Its commitment to artisanal luxury is connected through the creation of unique and personalized routes, built with the environmental and spiritual knowledge of indigenous communities and designers. These routes are created to recognize in each piece of luxury the human effort for beauty, refinement, innovation, purity, perfection, and experience [30].

Religare has participated in the most famous fashion fairs in the country, showing in each catwalk the articulated result of ancestral work and the thoughtful design, developed and processed carefully. Subsequently, the brand achieves to maintain tradition and cultural legacy in other ways of appreciating the subject premium, respecting natural assets, and manual labor at each stage of creation (Fig. 10).

The ancestral creative development linked to national brands such as Religare allows the country to know the Coreguaje community, its location, and its ability to work the Cumare, in other national settings such as parades, fairs, and exhibitions. These synergies allow the general public to recognize that in the Colombian territory, there are other latent and significant cultures for the historical development of the country and the fashion and artisanal luxury industry.

The most significant point about the articulation of this brand and the indigenous community they work with is the respect for the production times of the communities.

Fig. 10 Backpacks made of
Cumare fiber by the
Coreguaje community,
located in the department of
Caquetá, Colombia and dyed
with natural dyes (Image
sent by the designer Manuela
Peña, director and founder of
the Religare brand)

The slow production goes according to the rhythms established by the ancestral
peoples and the responsibility to propose new designs without altering the meaning
the Cumare has for the Coreguajes.

8 Fabric in Guanga

To the southwest of Colombia, in the department of Putumayo, is the Sibundoy Valley
and the Kamëntsá or Kamsá community and the Inga community. Since ancient
times, the agriculture, weaving, wood carving, the creation of musical instruments
and basketry have been the basis of the collective economy of both communities.

Each community has managed to distinguish itself through crafts, as a genuine
expression of their ancestral knowledge and an alternative of economic growth and
development amid a violent and vulnerable territorial context [10].

Although the Kamsá and the Ingas share cultural and organizational attributes, each has achieved through craftsmanship to differentiate themselves as a unique community, sharing their consciousness of the territory, its origin, language, and connection with the universe.

The Inga people, descendants of the Incas, are distributed in three departments of Colombia, concentrating more significant population in the Alto Putumayo and are known for their artisanal ability with weaving, knowledge of plants, and healing abilities.

Knitting is an ancient activity, created to cover the body and transfer oral history through intertwining, shapes, colors in a symmetrical game, where hands are responsible for translating visions, spiritual encounters, thoughts, past, present, and future desires and conceptions of permanence in the world.

"Telling stories and weaving are inseparable realities, irretrievably woven. For the Ingas, every story is first and foremost a fabric, a crossroads of stories, threads, and words that with some luck can become poetry. The art of knitting is analogous to that of cultivation. The sowers of stories are growers of the future: counting the past, they sing the present, and allow someone to listen and sing in the future."[13] Their knitting consists of shapes representing geometric figures and *zoomorfas*,[14] thus achieving the creation of ancestral crafts.

Wool is one of the primary raw materials used by this community, which in combination with plants such as elderflower, marigold, chilca, and yerbabuena, achieves diverse natural dyes. The weavers gather around the *"tulpa"* or homemade stoves to work with the fiber. During this process, they embody their connection with the land and theirs, through the *guanga*.

The guanga is a vertical loom used exclusively by the women weavers of the Inga and Kamsá community to make ruanas, blankets, scarves, and other pieces. The creations of these women are full of symbolism, which distinguishes their cultural and historical journey in the territory. This practice also manages to maintain the oral tradition through the practices of connection and familiarization generated by the art of knitting.

The fabric made in guanga can be labeled as an ancient art, which is transmitted from generation to generation through oral tradition. The making of these art forms maintains contact and balance with nature, as well as the symbology of the guanga, duality, relationships community hierarchy, iconography, and symbology of the people [10] (Fig. 11).

One of the most fabulous Ingas creations is the *Chumbe,* a sash or belt that is traditionally between 5 and 10 cm wide by 4 or 5 m long. Its central element is the geometric figure of the rhombus, an abstraction of the anatomical conformation of the stomach, which symbolizes the place where life begins; it is why women use it around the belly as a form of protection to the place where a new life is gestated.

---

[13] Accessed 1 July 2019 https://www.elespectador.com/noticias/cultura/benjamin-jacanamijoy-hijo-del-viento-articulo-713541.

[14] *Representations of animals through drawings and figures that express the importance of wildlife in the community cosmogony.*

Fig. 11 Cortizio skirt and
jacket in ivory by designer
Alvaro Leyton, made in
ancestral fabric by Doris
Jajoy (Image sent by the
designer Alvaro Leyton)

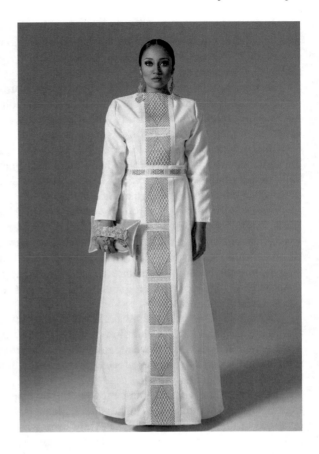

It is used horizontally, wrapped around the woman's belly and its manufacture is
done vertically; it is also the place of coexistence of men, the representation of the
four cardinal points that make up the world. The *chumbe* is also used to carry the
baby on his mother's back, creating a maternal bond that unifies and solidifies family
relationships and the mother-child-earth connection.[15]

The art of weaving these belts is called *chumbar* and means wrapping, hugging
and is performed by the hands of women who narrate symbolic stories conferred by
the gods throughout the universe and nature, loaded with knowledge, prosperity, and
abundance. "*Chumbar* is like knitting for the Ingas, inscribing a memory that resists
silence."[16] Through the chumbar they safeguard their legacy and traditional culture
by weaving the past, present, and future alive in the collective memory.

---

[15] Accessed 1 July 2019 https://coleccionetnograficaicanh.wordpress.com/el-chumbe/.

[16] Accessed 1 July 2019 https://www.elespectador.com/noticias/cultura/benjamin-jacanamijoy-
hijo-del-viento-articulo-713541.

Fig. 12 BIYA skirt and TAMO blouse by designer Alvaro Leyton. Crown of *chumbes* and the ancestral fabric of Doris Jajoy (Image sent by the designer Alvaro Leyton)

Doris Jajoy is a master Inga weaver, who has turned her art of knitting into a showcase of resilience, cultural development, resistance, and innovation within her territory. Through pieces that mix in each composition tradition, memory, and design, Jajoy is proposing a new way of perpetuating and safeguarding craftsmanship: fashion (see Fig. 12).

Jajoy, together with the designer Alvaro Leyton, managed to create a collection that revalues the fabric, as a form of collective unification that multiplies ancient knowledge and protects the earth as a source of inspiration and life. This collection exalts the cultural heritage rooted in indigenous peoples through colors and forms, creating pieces where luxury lies in all its forms of manifestation.

The correlation of discourses achieved between Coreguajes and Religare, Doris Jajoy and Álvaro Leyton, crafts and design, reconstructs and generates social and cultural fabrics and strengthens collective identities through material goods designed. These correlations also build collaboratively, narrating the life of a community, and its connection to the earth through their know-how, which is an invaluable art [40].

These new ways of exhibiting handicrafts and expressing their work on objects that manage to permeate the body and occupy essential spaces within the daily life of the being, build new ways of recognition and visibility of craftsmanship as a cultural and material expression of the Colombian territory. A need to safeguard each piece is born, in part due to the intrinsic history in its forms, the beauty of its colors, the quality of the materials, its environmental responsibility as well as the symbolic relationship that allows us to weave with our ancestors, with the origin, with who we are.

9 Conclusions

It is not possible to understand the design from Latin America if it is not empowered through its link with crafts and their differentiating factors in a globalized world. Subsequently, the need to articulate handicrafts to design is born, in its quest for resignification and reconstruction of a new luxury [40].

Traditional craftsmanship represents a unique mixture of skills with a more profound expression of human values that transcend culture, time, and space [34]. These values, articulated to design and new creative industries become exponential craftsmanship, where tradition, luxury, and emotional connection between producer–consumer converge, allowing parallel and appreciative dialogues to be established through an object.

For Hawley and Frater [34], luxury is defined as a high quality, limited quantity, creative and rare product/service and claimed that craftsmanship qualifies within these adjectives. If we go back to the first part of this chapter and articulate the different perspectives created around craftsmanship, we can affirm that each perspective converges on the fact that artisanal development as a material expression of a specific culture, has been from its origins a sustainable piece of luxury perpetuated in time through its multiple manifestations.

Campuzano [11] affirms by exposing the different conceptions of luxury that the basis and the rationale for the luxury product is the craftsmanship and its entire production process, from the way they extract the raw material to the exclusivity of the finished product.

Modern capitalism generates an extraordinary wealth, often at the cost of devouring natural resources more quickly than nature can replace them [35]. This phenomenon generates irreparable damage to natural ecosystems, altering the social balance visibly at the micro- and macro-level.

The creation of products based on capitalist guidelines begin to be outdated and the focus toward traditional concepts based on experience and the need to create sustainable spaces, where quality over quantity goes first, begins to take root as an alternative to transformation at the sociocultural, environmental, and economic levels [37].

Sustainability, luxury, and ethics initiate a permanent and unalterable dialogue in the reconfiguration of the fashion system. This dialogue allows addressing in a meaningful manner, the creation of pieces aligned with responsible processes allowing environmental and social balance in a society where luxury cannot exist without sustainability. In time, this mutual relationship between sustainability and luxury will become more robust, becoming the fundamental pillar of new creations [39].

As a form of cultural permanence and resistance over time, the artisanal development, and its broad cultural and immaterial weight, typical of the indigenous communities, has been in permanent dialogue with the earth as a sacred place, a space for spiritual and earthly connection and source of creative inspiration to narrate its legacy through colors, pints, materials, and know-how.

Handicrafts are personalized and exclusive products, which become small treasures in an ocean of homogenized products [32]. Crafts generate an emotional relationship between the craftsman and the consumer throughout history. On the contrary, mass production lacks the ability to create those relationships.

This direct symbolic interaction confers different experiential spaces, where all the senses get involved and give the acquisition, an experiential luxury dimension, where knowledge, the values exposed through the fabric, the connection established with the color through its natural process, the articulation of knowledge around the *tulpa*, the desires intertwined in each fiber and the mythical and ancestral narrative, gives each piece uniqueness, human value, and identity with history and culture.

Luxury craftsmanship, as a pillar of cultural sustainability in the construction of permanent social fabrics, generates spaces for collective collaboration, where different knowledge and creative perspectives converge, preserving intangible resources, essential for the continuity of indigenous communities [20, 21] and their know-how, as a material compilation of knowledge necessary to address design, sustainability, and luxury in the same space.

The designer–craftsmanship relationship promotes the creation of intercultural spaces and the exchange of technical and oral knowledge, for the creation of different proposals based on the recognition and dignification of artisanal work as the commitment of new cultural luxury immersed in societies of sustainable development.

Weighting the hierarchy of luxury on new conceptions associated with thoughtful purchases, the beauty of materials in the broadest sense, slow manufacturing, and respect for social and environmental aspects allows associating luxury with experiential and symbolic relationships. Relationships are established in the modern world through craft and explore other forms of socio-emotional bonding, which goes beyond the economic attribute.

Cultural appropriation changes from natural to synthetic materials that alter the intangible heritage of ethnic groups and the modification of handicrafts by various brands under the name of the peoples raised in the last decade, transgress the intangible wealth of indigenous communities. It also violates the individual, collective, and cultural rights of indigenous people, ignores intellectual property, and generates ruptures in the economic development of people.

> The market must understand that it is not two balls up or two balls down. These designs are images of their worldview. Communities ask for respect; they do not ask for money. They want designers to approach them and ask permission.

The development of public policies in favor of sustainable luxury articulated with cultural luxury and craftsmanship allows the promotion of sustainable models of life, thanks to its contribution to the generation of more balanced and fair societies.

Some other tangible examples of the capacity of artisanal luxury as a repairer of today's economic problems and a means of perpetuating craftsmanship as a form of cultural permanence are:

- Álvaro Leyton and Doris Jajoy and their union of design and ancestral fabric as a revaluation of artisanal aesthetics.

- Awana Taller is a platform of work and responsible dissemination of ancestral knowledge, meanings, and community representations of peoples.
- Bachué, a concept of handcrafted jewelry developed through knowledge and values shared with the Zenú and Wayúu community, adapting their pieces in the creation of jewelry keeping inscribed their original legacy.

Luxury brands nowadays spend more money on marketing and selling that in designing and developing the product. These communication expenses, of course, are a response to an interconnected world that makes it necessary to hire influencers, magazines, social media ads, and much more. Today, the name of a brand can be even more important than the quality of its products [25, 26].

Knowing that this is how the system works, it cannot be possible that brands do not use that massive reach to educate on the stories behind the pieces [44]. They will certainly require to know first the meaning behind it, and for that, they need to work with the communities, take the time to understand and let them instruct the right ways to use.

Both uniqueness and durability are essential characteristics of luxury. They are also intrinsic values of craftsmanship. Enabling communities to offer their crafts as elements of luxe, supporting their production processes, and enhancing their business capacities is, without doubt, a way to also promote sustainability in the fashion industry.

However, it is essential to understand that for artisan communities, crafts mean much more than a label of a luxury item. It means stories told differently. It means craftsmanship qualities that have been passed on from generations from the elders who passed on their wisdom through music, art, and rituals.

Finally, understanding craftsmanship from a slow, sustainable, ethical, exclusive, and high-quality approach, it would be possible to think:

Is craftsmanship an intrinsic part of luxury, or indeed, is luxury a characteristic immersed in craftsmanship?

References

1. Acosta E (1991) Artesanías. Nueva revista colombiana de folclor 3(11):81–88
2. Aguilar M (ed) (2010) Las artesanías en el contexto global. Artesanías de América, Centro Interamericano de Artesanías y Artes Populares 70(2):67–90
3. Angelotti G (2004) Artesanía prohibida: de cómo lo tradicional se convierte en clandestino. Colegio de Michoacán, México
4. Appadurai A (2001) La modernidad desbordada: dimensiones culturales de la globalización. Fondo de Cultura Económica, Buenos Aires
5. Armando M (2016) Lujo sustentable. In: Essentia consulting. http://www.essentiaconsulting. net/lujo-sustentable/. Accessed 8 June 2019
6. ARTESANÍAS DE COLOMBIA. www.artesaniasdecolombia.com. Accessed 15 June 2019
7. Asociación de artesanos y artesanas (2005) Pintando nuestra cultura Zenú. San Andrés de Sotavento

8. Baracich C (2016) Artesanías: territorios de construcción de "saberes otros". Artesanías de América, Centro Interamericano de Artesanías y Artes Populares 75(1):29–37
9. Barrera E (1997) El diseño precolombino, un espacio para la inspiración. Consejo Editorial de Autores Boyacenses Editorial, Boyacá, Colombia
10. Caipe R, Hernández M (2013) La dualidad andina del pueblo pasto, principio filosófico ancestral inmerso en el tejido guanga y la espiritualidad. Plumilla Educativa 11(1):136–156
11. Campuzano S (2016) La fórmula del lujo: Creación de marcas, productos y servicios. LID Editorial
12. Canclini N (1990) Culturas híbridas. Grijalbo, México
13. Castro I (2003) Reflexiones en torno a la artesanía y el diseño en Colombia. CEJA Editorial, Bogotá
14. Chamorro I (2006) Artesanías y cooperación en América Latina. Centro Interamericano de Artesanías y Artes Populares, Cuenca-Ecuador
15. Delgado M, Gardetti M (2018) Vestir un mundo sostenible: La moda de ser humanos en una industria polémica. LID Editorial, Argentina
16. Entwistle J (2002) El cuerpo y la moda; una visión sociológica. Paidós, Barcelona
17. Fletcher K (2008) Sustainable fashion and textiles, design journeys. Earthscan, London
18. Fletcher K, Grose L (2012) Fashion and sustainability: design for change. Laurence King Publishing, London
19. Gardetti M (2011) El sector del lujo y los valores de la sustentabilidad. Centro de Estudios para el Lujo Sustentable, Buenos Aires
20. Gardetti M (2018) Textiles y Moda: ¿Qué es ser sustentable?. LID Editorial, Argentina
21. Gardetti M (2018) Lujo Sostenible. LID Editorial, Argentina
22. Gardetti M, Girón M (2014) Sustainable luxury and social entrepreneurship: stories from the pioneers. Greenleaf Publishing, Sheffield
23. Gardetti M, Torres A (2011) Moda y diseños lentos. Publicación Colombiana de Tecnología y Educación 16:9–13
24. Gardetti M, Muthu S (eds) (2016) Ethnic fashion: environmental footprints and eco-design of products and processes. Springer
25. Gardetti M, Muthu S (eds) (2016) Green fashion, vol 1, Springer
26. Gardetti M, Muthu S (eds) (2016) Green fashion, vol 2, Springer
27. Gardetti M, Muthu S (2016) Sustainable luxury, entrepreneurship and innovation. Springer
28. Gardetti M (2017): Sustainable management of luxury. Springer
29. Gardetti M (2015) Textiles y moda: ¿qué es ser sustentable? LID Editorial, Argentina
30. Girón M (2015) Forward. In: Gardetti MA, Torres AL (eds) Sustainable luxury: managing social and environmental performance in iconic brands. Greenleaf Publishing, Sheffield
31. Godart F, Seong S (2014) Is sustainable fashion possible? In: Gardetti M, Torres L (eds) Sustainability luxury: managing social and environmental performance in iconic brands. Greenleaf Publishing, Sheffield
32. Guldager S (2015) Irreplaceable luxury garments. Springer
33. Hashmi G, Muff K (2015) Evolving towards truly sustainable hotels through a "well-being" lens: the S-WELL sustainability grid. In: Gardetti M, Torres L (eds) Sustainability in hospitality: how innovative hotels are transforming the industry. Greenleaf Publishing, Sheffield
34. Hawley JM, Frater J (2015) Economic impact of textile and apparel recycling. In: Sustainable fashion: What's next?, 2nd edn. Bloomsbury, New York
35. Honoré C (2006) Elogio a la lentitud: un movimiento mundial desafía el culto a la velocidad. RBA
36. Isla A (2009) Los usos políticos de la identidad: criollos, indígenas y Estado. Libros de la Araucaria, Buenos Aires
37. Korten D (2009) Agenda for a new economy: from phantom wealth to real wealth. Berret-Koehler, San Francisco
38. Maglia G, Hernández L (2017) Memorias, saberes y redes de las culturas populares en América Latina. Universidad Externado de Colombia, Bogotá
39. Maisonrouge (2013) The luxury alchemist. Assouline Publishing, New York

40. Malo G (2015) Tejiendo relaciones significativas entre nudos, urdimbres, tramas y texturas. Artesanías de América, Centro interamericano de artesanías y artes populares 74:35–42
41. Marquéz E (1997) El diseño precolombino, un espacio para la inspiración. Tunja, Boyacá
42. Max Weber (1996) Economía y sociedad. Fondo de Cultura Económica, México
43. Ministerio de cultura colombiana (2003) Impacto económico de las industrias culturales en Colombia. Bogotá (in press)
44. Muthu S (2018) Sustainable fashion: consumer awareness and education. Springer
45. Organización nacional indígena de Colombia. https://www.onic.org.co. Accessed 23 Jul 2019
46. Orígenes (2014) Honrando las raíces de la artesanía colombiana. Ecopetrol y Artesanías de Colombia, Bogotá
47. Puche B, Suárez M (1991) El sombrero vueltiao Zenú, el tejido de la caña flecha. Nueva revista colombiana de folclor 3(11):89–115
48. Quiñones A (2003) Reflexiones en torno a la artesanía y el diseño en Colombia. Universidad Javeriana, Bogotá
49. Rodríguez E (1991) Artesanías. Nueva revista colombiana de folclor 11(3):81–88
50. Rodríguez M (2017) Mujeres I´ku: tejiendo otras formas de permanencia. Universidad de los Andes, Bogotá
51. Sandoval A, García L (2013) Artesanías, cultura y desarrollo. Artesanías de América, Centro Interamericano de Artesanías y Artes Populares 73:30–35
52. Saulquin S (2014) Política de las apariencias: nueva significación del vestir en el contexto contemporáneo. Paidós, Buenos Aires
53. Serrano J (2015) Descubriendo la moda ética con Stella Jean. Revista Forum de Comercio Internacional 1:28–29
54. Serrano J (2015) Artesanía y Globalización. Artesanías de América, centro interamericano de artesanías y artes populares 74:58–67
55. Vega D (2013) El Campo Artesanal: Aporte teórico social y pedagógico. Fundación Universitaria Juan de Castellanos, Bogotá
56. Vézina B (2019) Curbing cultural appropriation in the fashion industry. CIGI Papers, Canadá

Alejandra Ospina is Foreign Affairs Professional and Political Scientist of the UMNG. Fashion Studies at Parsons School of design. Expert on Social Enterprises and Sustainable Fashion. Magister in Corporate Communication Management at the University of Barcelona. Founder and Director at TRIBECO and co-founder at TRIBECO Foundation.

Ana López is Sociologist of the University of Antioquia specialized in Sociology of the Design of the University of Buenos Aires. Studies on Sustainability of Fashion of the London College of Fashion. Director at WALIRUU and Project Manager at TRIBECO.

Crafting Luxury with 'More-ish' Qualities at the YSL Museum: An Organic Approach

Annette Condello

Abstract Today, what is organic within architecture enables local craftspeople to acquire new expertise. Studio KO's Yves Saint Laurent Museum (2017) in Marrakesh, Morocco, is imperative for understanding the relation between sustainable luxury and local craftsmanship. The construction fortifies fashion with the environment as well as local brickwork traditions. For the architects, horizontal layers of textured brickwork resemble a textile weave and become a constructed cultural entity. This chapter traces the 'more-ish' qualities, defined herein as causing one's impulse to create *handsome* foci (as opposed to the Moorish, which is culturally based), inherent in the YSL Museum from an art-architectural perspective within the sustainable luxury context. It explores the display of the camouflaged Berber (or *Amazigh*) textile within the building's facades and interprets its 'more-ish' qualities of luxury through its spaces. Analysing Howard Risatti's 'theory of craft' [42] and Richard Sennett's idea of craft 'as an enduring, basic human impulse' [21], this chapter explains why the bricks-and-mortar structure and *zellige* (hand-made) tiles have literally and metaphorically regained value through craftsmanship in the luxury sector. It discusses how traditional brick-laying techniques and their unexpected connections have transformed the desert built environment and speculates why these changes inform adaptive reuse practice as a 'more-ish' organic approach. In this respect, in discussing the brickmaking as a form of crafting luxury the process has cultivated a bonding or 'tuning-in' tactic, important in understanding sustainable Moroccan culture.

Keywords Organic architecture · Sustainable luxury · Berber textiles · Studio KO · Brick technologies · Recrafting *Amazigh* culture

1 Introduction

'Fashion was a craft, a poetic craft'

[41, p. 20].

A. Condello (✉)
School of Design and the Built Environment, Curtin University, Perth, Australia
e-mail: a.condello@curtin.edu.au

© Springer Nature Singapore Pte Ltd. 2020
M. Á. Gardetti and I. Coste-Manière (eds.), *Sustainable Luxury and Craftsmanship*,
Environmental Footprints and Eco-design of Products and Processes,
https://doi.org/10.1007/978-981-15-3769-1_7

Marrakech was, and still is, recognised primarily as a tourist destination for non-conformists whom made, or make, the extra effort to bond with ancient cultures and popularise Morocco's unfamiliar luxury crafts abroad. Marrakech's location was considered the 'cross-roads of the whole north-west of the continent … where desert caravans arrive and depart when Bedouin meets Moor and camel meets lorry' [23, p. 178]. After Morocco's Independence (1956) as well as the counterculture of the 1960s and 1970s, French haute couture designer Yves Saint Laurent frequented the place and collected an extensive range of Berber luxury goods (Fig. 1). Eventually, Saint Laurent made his atelier and home there. Morocco's private economy sector grew in the 1980s. And by the mid-1990s, Marrakech's entrepreneurs asserted how 'the city's proximity to the main European capitals, promoted it as a convenient-yet-exotic-alternative to host conventions and small-scale events' [22]. Luxurious hotels and resorts were erected and 'a class of new hoteliers and investors sought to broaden the "luxury" niche market and reach out to those who could afford it' [22]. Marrakech became recognised as an enchanted city because of its exotic desert landscape and luxurious accoutrements and in the context of modern and contemporary craftsman-ship [34] and their links with the fashion industry. Morocco offers more insight into the evolving forms of forgotten crafts and their connections to international trade beneath and on the earth, and in the atmosphere than what one might expect.

Comprising Berber tribes, arising out of the desert in the eleventh century, when Yusuf ibn Tashufin established Marrakech as the capital of the Almoravid Empire,

Fig. 1 Berber carpet-seller with a photo of Yves Saint Laurent on the wall, the place where he use to buy his rugs, at Marrakech's souk. Photo courtesy: L. Broscatean, 2019

the city developed into a permanent strong-hold. Berbers deserted their itinerant tents [32] and mud homes. It was Abu Bakr, leader of the Almoravids, who built the first *kasbah* (stone house) in Marrakech for the desert Berbers [7, p. 467]. The first attempt at making it into a sustainable desert-city in Morocco that linked the ancestral place in the Atlas Mountains was through the instalment of underground irrigation channels (*khettara*) [13]. Water could filtrate through a series of rills to the buildings and surrounding gardens within and outside the medina [13]. Today, what remains in Marrakech's medina, a populated area with its relaxed geometric city-grid (and now a UNESCO World Heritage Site), are masonry buildings punctuated with luxury boutique riads, or courtyards, and some dotted with reflective pools. Above and below ground, the network of *khettara* acts as a figurative and transient 'Berber loom' with its straight warp in stationary tension and moveable liquid weft. 'Each object from the past is like a silent word' [17, p. 385].

In 'Looking at Mars in Marrakech', Abdelkader Benali writes:

Marcel could see Marrakech crystal clear from the air. It was a pleasure to look at. It was only later that he understood that the transparency was deceptive. It was difficult not to be overwhelmed. The city had proved itself resistant to the tourism hype. It had not only survived it, but had given it a twist …

She didn't walk towards him, she floated, as if a gigantic wind turbine blew her along. The refinement that hid her dark past as if it were a secret weapon had become even more intense.

'We have seen Mars. He's been on the red planet' [6, pp. 65–74].

The above quotation from Benali positions this discussion in tracing the origins of Marrakech's museum for the fashion mogul Yves Saint Laurent within the realm of sustainable luxury and craft (Figs. 2 and 3). I use the term 'more-ish' qualities in this chapter to refer to one's impulse to create *handsome* foci (as opposed to the Moorish, which is culturally based) concerning crafting luxury-layers of history that have been un-earthed which also provide a new authenticity for the potential for reuse.

Owing to the physical evidence of meteorites crashing through space towards Earth, fallen rust-coloured objects from the past are popular for international trade, especially in Morocco—stones move around by hand on the market. Black-and-red meteorites from planet Mars, for example, are found amongst windswept sand and preserved in Morocco. Curiously, in the early 1980s, a clump of meteorites 'of Martian origin represented a breakthrough in attempts to understand the geological evolution of Mars' [3, p. 785]. After their arrival on Earth, however, the collected samples 'experienced variable degrees of terrestrial weathering' since they were 'exposed to organic and other potential contaminants during storage' [3, p. 785]. In some ways, these space-junk samples are recyclable in the sense that they formed new rock types. Left bare and to be re-found, this rock-type might have served as the lasting foundation rock for the built-up ramparts in Marrakech in the Middle Ages, made of a distinct pink-and-red chalk-and-clay, earning the city's reputation as the 'red city'. As space-junk [24], the meteorites themselves govern more than the ramparts through its earthbound permanency. 'If space-junk is the human debris that

Fig. 2 YSL Museum's
open-air drum courtyard.
Photo courtesy: F. Tiaiba,
2017 ©

litters the universe, junkspace is the residue mankind leaves on the planet …. More
and more, more is more', which is now occupied by 'a look-no-hands world' [25].

In Marrakech, more hand- and machine-crafted spaces co-exist where the soil and
architecture materialise as one, rust-red, and from the macro to the micro-fibre scale.
In particular, the textural qualities of brick-craft, impart organic qualities through its
variegating building facades (Figs. 4 and 5). What was organic within architecture
enabled how local craftspeople could adapt and improvise new skills, such as the
intricate weaving patterns evident within brickmaking and Berber rugs. When considering both the craft and fashion industry in Marrakech, 'all textiles begin with a
twist' [21, p. 4].

2 Twisting Yves Saint Laurent's Mark in Marrakech

Today, Marrakech is increasingly becoming a tourist-Mecca for Saint Laurent fash-
ionistas. With its distinctive red-coloured brick-pelts applied to the building's upper
facades, the contemporary twisting of terracotta textiles embedded within a Marrakech museum exists near the French Art Deco-inspired Gueliz district. In Morocco,

Fig. 3 YSL museum's brick
and terrazzo façade. Photo:
A. Condello, 2019

'the French changed the urban structure of the country and introduced a modern approach to architecture and heritage management' [27, p. 136]. The aim was to 'support the colonial decentralization of the political and economic power by shifting it from the pre-colonial capital, which generally switched between Fez and Marrakech … to the new coastal urban centres' [27, p. 137].

Marrakech's riads, or courtyard houses, nuzzled between the formal and haphazard layout of its medina (*kasbah*) informing Moroccan crafts, presented new ways to look at how planning contributed to the landscape as a luxury attribute within its architecture. In 1913, French architect Henri Prost designed Marrakech's first masterplan [27, p. 137]. 'In contrast to the medina's organic layout,' Prost 'designed the *ville nouvelle* concentric and linear. While the medina's (or pre-colonial city) landscape emphasised morphological unity and the visibility of specific landmarks such as mosque minarets, the *ville nouvelle's* modern planning was based on granting straight transportation axes, monumentalizing of state buildings and zoning policies' [27, p. 138]. By the 1950s the country had changed incrementally. For Kidder-Smith, 'Morocco, both in architecture and in customs, is the most remote and oriental region in North Africa. It has the highest mountains, is the least explored, and was the last pacified. It is also the richest and fastest-growing' (1955, p. 177). Intriguingly, according to Lamzah, Gwendolyn Wright argues that the French 'museumified Moroccan

medinas' [27, p. 3] and it was this aspiration that enlightens the Yves Saint Laurent
museum's well-crafted design (Fig. 6). The city of Marrakech thus serves as a point
of departure for analysing the refinement, transparency yet steadfast brickwork.

Yves Saint Laurent Museum (2017) continues to advance the link between aspects
of the atmospheric lucidities and simplicities of local craftsmanship within the constructed landscape. Intriguingly, Gueliz was the first district to be constructed outside
the Marrakech medina with views through to the Atlas Mountains. There, a building code was introduced which indicates that no structure could be erected at the
same height of a palm tree (equivalent to three storeys) (Fig. 7). The location of the
neighbouring Jardine Majorelle was informed by two preceding luxury attributes:
the aesthetics and views through to the Atlas Mountains were already affixed to the
district.

In the 1980s, French industrial patron Pierre Berge (1930–2017) and his business
and life partner Yves Saint Laurent (1936–2008) saved the neighbouring Jardine
Majorelle from ruin. They bought 1930s modernist house and garden belonging to
French artist Jacques Majorelle. Following Saint Laurent's death in 2008, three years
later Berger established the Berber Museum located in the former painting studio
of Majorelle, an extensive collection of textiles, bold silver and gem jewellery and
ancient terracotta bee-combs set within a Saint Laurent type-of dark space universe,

Fig. 5 Traditional
brickwork at Jardine
Majorelle, Marrakech.
Photo: A. Condello, 2019

by both Berge and Saint Laurent. Adjacent to the Berber Museum, in the same block, is the Yves Saint Laurent museum which is characterised by its thin concrete mantle, pink-coloured terrazzo walls above which are five different types of red brick facades attached with swathes of traditional brickwork patterns (Fig. 8). Berge made 'the effort to make Saint Laurent part of the cultural heritage of the nation' [33], however, the building itself tells a new story—one demonstrating its North African poetic craft-connections.

Crafts embedded with utilitarian uses are attractive through camouflage and sustainable through their Moorish origins, as in the case of the Yves Saint Laurent Museum, designed by Studio KO office (Karl Fournier and Olivier Marty), and located along the renamed street: rue Yves Saint Laurent. It is an homage to the haute couture designer's work and legacy [29]. This haute couture museum arouses and educates the viewer to better understand the crafting of fashion within the sustainable construct. It certainly has created Morocco's most crafted luxurious landmark.

Considered by some as one of the most sustainable places on the globe, Morocco 'is a pioneer among the Middle East and North Africa (MENA) countries in establishing a policy and regulatory framework for promoting renewable energies and energy efficiency' [26, p. 55]. The brickmaking sector, however, is problematic. Generally

speaking, Morocco produces about five million tonnes of clay bricks per year. Brick-making impacts one's health and the environment it affects the air quality. Part of the solution to this problem is that brick-building weaving techniques are changing contemporary architecture in Morocco as a sure-fired sharp practice to improve sustainable luxury craftsmanship. Fortunately, 'craftsmanship is Morocco's future' [35, p. 30]. If products are man-made rather than machine-made, then clay-fired brick products have become more sustainable in today's sense of the popular-reuse properties of organic products. And presents unexpected approaches to incorporating brickwork within contemporary architecture in innovative ways.

3 Crafting Luxury, Berber Textiles and Brick Making

Today, when we think about rampart cross-cultural consumerism and globalisation of an object as well as their link to mindful craft, much emphasis is placed upon on where a luxury product comes from and where it is made, before it is sold online. It is rarely considered in relation to historical change or the actual archiving of physical luxury products such as drawings, clothing and accessories or its display in the crafted

Fig. 7 YSL museum construction site. Photo courtesy: F. Tiaiba, 2017 ©

museum. The display of traditional luxury objects is what makes long-lived crafted forms more satisfying and educational in its original cultural context. For instance, in Northern Africa, for instance, 'the loom is widely regarded as a powerful, living creature' [28, p. 44].

In philosophical terms, luxury has metamorphosed from something material (and living) to something now considered an aesthetic of experience. For Lambert Wiesing, 'luxury always develops from a thing with a purpose. That is why luxury is experienced only in that use which is possible for the possessor, that is, only when a luxurious thing is used for the purpose it serves with irrational effort in a nonpurposeful or unpurposeful way' [44, p. 136]. 'Because what is the case for luxury is also what defines art …' [44, p. 136]. Craft is considered herein as excessive, 'more-ish' but a useful art. Craft itself is also luxury for all to experience and relish in the visual indulgence of perfections and imperfections to be encourage others to experience satiation.

Before the age of industrialisation, most, if not all, trades and creators crafted for utilitarian purpose, often ornamented. To some extent, the true sense of manufacturing a 'crafted' luxury product has partly lost its meaning. Made somewhere and made by hand is what people now associate with craft and what a high-quality luxury product is. Irrespective of the actual crafting process, what has dominated the luxury market is people's desire to pay for a bespoke item which is hand-made, machine-made or machine-made to look as though it is hand-made with grit, sometimes bordering on the absurd, such as products made purposely to look as though they were made by hand, but in fact made by a machine. More than the essence of

Fig. 8 YSL museum
showing three different types
of brickwork patterns. Photo:
A. Condello, 2019

the product of time, the value of experience appears through craft and the aesthetic of experience is now counted or accepted as a new measure of sustainable luxury. Why reverse the painstaking process?

And yet how did time become the essence of handcrafted luxury products? Apart from the shifting sands in an hourglass, referring to an ethnographic example in eighteenth-century Morocco, 'we know that Europeans offered gold and silver watches to various dignitaries of the Moroccan "courts" and also that they believed that Moroccans used these artefacts as body ornaments or objects of prestige' [1, p. 37]. Together with the handcrafted imported European luxury products, by the 1920s there was a growing demand for quality clocks in Marrakech as they were highly appreciated [1]. As far as time-shifting and the time it takes to produce high-quality luxury goods is concerned, not much has changed since then. The mindful crafting of watches still presents the ultimate form of high-quality luxury craftsmanship. Digital media, however, has altered the advertising campaigns of craftsmanship, where one is able to view online any crafts instantly, which extends to benefit the architectural practice to sustainably craft luxury.

In terms of craft itself and craftsmanship, 'sustainable luxury' is concerned with, amongst other things, 'preserving the cultural heritage of different nations' [16, p. 55]. When rethinking the crux of sustainable luxury, its connection to craft and fashion as well as its association with what constitutes organic architecture, preservation alone of architecture is substantial enough.

At the core of craft, traditional brickwork, especially the time it takes for layering techniques emanating from ancient Berber culture as with its textiles and stepped-pyramidal remnants in the Sahara desert, North Africa, have a profound underlying presence in Marrakech's architecture. Running bonds, twisted and crossed bricks are most certainly Berber in origin and earlier abstract techniques used in Mesopotamia. Outlining the original presence of brick in cities as the oldest hand-made material, Dan Cruickshank states that in Urak, Mesopotamia, founded 6000 years ago created sun- and kiln-fired bricks in building the city, such as the Ziggurats of Eanna and at Ur. Cruickshank then states how in *The Epic of Gilgamesh*, the King of Urak 'sought immortality and found it' [10]. King Urak 'realised that his name stamped on the hard bricks, "where the names of famous men are written" meant that his creations and his memory would last for eternity. The kiln-dried brick was the passport to immortality, a guarantee that your creations—and your names—would live forever' [10]. As far as brick manufacture is concerned, The Book of Exodus recounts 'when the "Children of Israel" were exiled in Egypt the Pharoah increased their labour by refusing to give them straw to make their bricks (Exodus 5: 12–14) ... straw was mixed with clay, which would bind sun-dried bricks and help as fuel if the bricks were kiln-fired ... Arguably brick making has much to do with the subsequent history of the world' (cited in [10]).

Through time, brick technologies improved and helped strengthen walls and transfer loads to create intricate bonding patterns and functional luxurious spaces. 'In this age of ever-increasing concern over ecology, sustainability and energy conservation, bricks—with their long-life span and splendid insulation characteristics—remain an ideal building material' [10]. The poise and extrusion of the individual brick-bands we find within its traditional brick-laying techniques have thus transformed the desert built environment profoundly and aesthetically, especially within Studio KO's work.

Apart from the sustainable and textural qualities of brick-craft are the Berber textiles themselves. In *Berber Costumes of Morocco: Traditional Patterns*, Marie-Rose Rabate and Frieda Sorber discuss how 'the structure and textures of garments in themselves are powerful markers of Berber identity ... [And] the woven part of the Berber identity may well have its roots in antiquity. The textures of many textiles from Moroccan looms strongly resemble archaeological finds from ancient Egypt' [28, p. 25]. In North Africa, the loom 'is reborn every time a new warp turns loose beams and sticks into an entity, which empowers women to weave the fabrics that provide shelter, protection and cultural identity for their community' [28, p. 44]. The Berber loom clearly imparts links to architecture through constructing static textiles. Rabate and Frieda Sorber continue:

> Berbers apply designs to functional objects...or ceilings. Designs protect the objects that are necessary for the basic functions of domestic life. The evident decorative function of patterned garments and jewellery was secondary to the hidden, beneficial influence that had led to their adoption in ancient times. The makers of these objects were only known in their own group. 'Art for art's sake' did not exist among the Berbers [28, p. 272].

Berbers were more interested in living an authentic life through luxury craftsmanship. More importantly, 'rather than calling themselves "Berbers", a pejorative term derived from the Latin word *barbarous* or "barbarian", they refer to themselves by the name of the particular group. Berbers also use the overarching term "Imazighen" or "the free people"' [5]. It is precisely this quality that Studio KO architects adapt to include the horizontal layers of textured brickwork, resembling a Berber (*Amazigh*) textile loom in operation camouflaged in a biomorphic manner into what I call permanent brick-pelts, or a conglomerate of intricate brick patterns made to resemble snakeskin or date palm tree trunks.

Studio KO architects transform the textile through the constructed technique of weaving the architectural-cultural object. Amazigh craft is transcended into Morocco's postmodernist and contemporary architectural building techniques. In terms of Postmodernist attitudes towards craft production, Andrea Branzi states that 'the supposed creativity of the artisan, or at least his [or her] "ancient wisdom," is contrasted with today's consumption-oriented technology. In reality, the arts and crafts, shifting the responsibility for research into new forms of merchandise and consumption onto industry, stick to the pure and simple "reproduction" of existing models, i.e. those already shaped by tradition' [8, p. 580]. Amazigh tradition is what drives Studio KO architects' craft method as an authenticated- and luxury-quality expression.

In *A Theory of Craft: Function and Aesthetic Expression*, Howard Risatti notes that 'material and technique go to the very heart of craft' and is associated with strength, force, power, virtue' [40]. Interestingly enough, craft accentuates the 'technical knowledge and technical skill required to make an actual object come into being' [40], as if with the *sleight-of-hand*. For Risatti, 'skill of this kind was so useful and so extraordinary that in the Middle Ages the word craft also became associated with "witchcraft", a vestige of which remains in our use of 'crafty' for a shrewd or even underhanded person' [40]. In terms of brickwork, the sleight-of-hand-work and the clay brick material were forged together via the kiln as if cooked until burnt. The furnace for the bricks created the desired material poise of an object, as if rendered by sophisticated magic for the creation of new terracotta patterns or brick trickery.

In connecting sustainable luxury and craftsmanship, the tradition of incorporating brick trickery linked with the craftsperson's sleight-of-hand-work is an intriguing cultural after-effect through contemporary architectural practices. As already noted, the production of Studio KO's work transcends Amazigh culture. Traditionally, Moroccan women who practise craftsmanship (recognised as *Maalems*) are responsible for making intricate Amazigh carpets. Aspects deriving from their textile techniques have been adapted by Studio KO as evidenced in their brick façade compositions. The actual bricks were made in the Tetouan area, central Morocco.

Another technique Studio KO includes is the medieval tradition of making the monochrome green-glazed ceramic and tiles called *Tamegroute,* also the name of town east of Morocco's Sahara desert. These hand-made techniques made out of clay dug out from palm groves were brought to Tamegroute by the artisans from Fez. Aspects deriving from these ancestral-cultural prospects are repositioned abstractly within the contemporary architectural practice as with the application of *zellige* or mosaic tiles, which show deliberate and some imperfections—or the perfected hand-made markings. As master craftsman, Studio KO incorporated this luxurious artisan technique by inserting the green zellige tiles within the museum's atrium (Fig. 9), perhaps as a homage to Tamegroute itself to expose its lost cultural horizons.

Craftsmanship, for Risatti, 'is a way of making that provides a universal standard for judging objects and work itself that transcends cultural horizons because, as a process connected to the hand working of material, it is available to everyone and thus occurs worldwide' [40]. In hand-making products, appreciation of the effort and skill makes one question the origins of specific ancient practices. Originally influenced by Carthage, some scholars consider Amazigh culture as older than the luxury practices of the Egyptians, who were, amongst other things, experts in mummification and preservation of precious objects and foodstuffs for the afterlife. Craftsmanship thus affected our appreciation of the idea of the organic within architecture as a form of sustainable luxury. Intriguingly, it also engages ideas relating to the uncommon mysteries of craft.

In terms of what is sustainable today, sociologist Richard Sennett's *The Craftsman* suggests that it veers towards 'living more at one with nature' [42]. This has always been the case and Amazigh culture represents this, especially when observing the techniques used by the ancient Egyptians in context. He argues, 'the invention of the

Fig. 9 Museum's *zellige* laid within the atrium, paying tribute to *Tamegroute's* traditional green-ceramics technique. Photo: A. Condello, 2019

fired brick about 3500 BCE marked a turning point in brick construction, the bricks now strong in all seasons, serviceable in a variety of climates' [42]. Moreover, 'the invention of the fired brick was inseparable from the invention of the oven; some evidence suggests that the same enclosures were first used for both cooking and constructing. In cooking bricks, the kiln walls do work no open-air fire could' [42]. As a result, the firing process enabled bricks to be crafted well, which led to the subsequent invention of the 'frogging machine' contraption.

Baking bricks and cooking food in the same kiln are inextricably linked as a dual craft, saving energy for one purpose. In rethinking about what was organic then and how this relates to Morocco today, in 'Architecture and food composition' Peter Kubelka writes, 'bricks baked in a kiln had to be invented. By doing this, man had created a valuable new material that made it possible to build much more complicated buildings than simple huts to live in. Building with fired bricks revolutionised the practical possibilities of architecture, and with the architects' imaginations' (2007, p. 17). And since bricks are handcrafted components, 'they are formally repetitive and can be used to create much freer shapes than elements adopted directly from nature like stone blocks or tree trunks' (2007, p. 17). Cooking food and baking bricks and their improvisation are inextricably linked since the kiln's function was twofold: it exposed a new technique and demonstrates how 'more-ish' qualities might have transformed from biomimicry to formulate lightweight crafted bricks.

This was particularly the case in the nineteenth century, after the Franco-Moroccan war, with the advent of the Industrial Revolution. (And previously in the twelfth century, with the increase of the Silk Route in Northern Africa initiated by Moroccan traveller Ibn Battuta—between China in the East and Scandinavia in the North—accelerating the manufacturing of quality luxury products.) 'With the enhancement of the quality and appearance of brickwork and with the increase in production and the reduction of cost, the popularity of brick as a building material grew' [38, p. 53], especially in the late nineteenth century. Bricks were often made at the construction site. Brickmaking patents were invented and issued (as in the case of the USA and elsewhere), and the processes of firing and making moulds became more advanced. 'Brick masons desired efficiency in the use of materials and labour as well as bricks with higher quality' [38, p. 54]. Cast iron contraptions for moulding bricks were invented as 'frogging machines' whereby '*kicks* for forming *frogs* (indentations in the bottom of bricks) were built into the moulds' [38, p. 54].

Curiously, ancient Egyptian brick makers had improvised a crafting technique to save more clay when firing bricks by interring live 'frogs' in the brick. Hence, '"frogging"—the familiar (usually pyramidal) indentation in bricks—originates from the ancient Egyptian custom of creating hollows in their Nile-clay bricks, by interring live animals, but some historians dispute this deplorable technique (quoted in [39]). Assisting the pharaohs, the myth of *Heket*, an Egyptian Frog goddess was invoked in magical charms to protect the dead on their way to the afterlife as well as defending the home. Returning to the form of the lightweight brick, the 'frog recess' hence made it easier for the local craftsman to make more bricks with the excess clay and make them lighter to lay, which also reduced the amount of time to char the bricks in kilns. 'Traditional Egyptologists favoured the after-life scenario (baby animals ready to grow to serve the risen Pharaoh), until the 1903 discovery of millions of skeletons of "Bufo regularis"—the common African frog—in the remains of ancient Egyptian workers' buildings on the Giza Plateau' (quoted in [39]). 'Although this amphibious exhumation' and connection to brickmaking 'was not well known other than by historians and palaeontologists, the Victorian trades embraced this romantic [sic] custom during the industrial revolution, and it's been with us ever since, albeit buried' (quoted in [39]). At any rate, 'frogs were a significant improvement in the moulding, firing and laying of bricks, for several reasons. They saved a significant amount of clay, reducing the intensive labour of excavating and preparing clay, and lessened the weight of bricks for easier transport. They allowed for faster drying of green bricks before firing and promoted better internal firing of bricks ... frogs could also help in making taller bricks; thereby improving bonding' [38, pp. 54–55].

By the early twentieth century, however, 'frogging machine' contraptions and portable brick presses became outmoded. Whichever craft process was acceptable at that time, live or artificial in-fills, have most definitely altered the 'frog' bricks during industrialisation. This is especially the case in Britain when patented 'frogging machines' led to the manufacturing of hollowed shaped bricks so builders could place their markings on the indentation of the brick. Initially, an insensitive trade, the making and branding of bricks was deceptive but became more transparent. Consequently, the brickmaking trade was replaced by a bewitched craft, which was supposedly considered more genuine as it was guarded by its indented markings.

Sennett believes solid bricks were 'honest' in the sense that:

all the bricks laid, say, in a Flemish bond course comes from the same kiln, and even more, 'honest' brick evokes a building surface in which the brickwork is exposed rather than covered over; no cosmetics, no 'pots of whore's rouge' have been applied to its face. One reason for this shift was that masons were beginning to be aware of, and feel engaged in, debates about the meaning of naturalness as opposed to artifice—the great Enlightenment preoccupation about nature brought home to the proper use of a natural material.

The work these metaphors did on bricks can be understood through the attitudes we now harbour about organic food. Strictly speaking, organic food refers to purity of substance and minimal manipulation in production [42].

Honest bricks though are not always authentic. It all depends on their material-ity make-up. Glazes alter the natural process. Nonetheless, the thought of charring bricks with real frogs defied nature. This natural-artificial technique did not render an organic or sustainable practice, but there is an important case for repositioning how the connection to food and its transformation within architectural compositions. 'Improvisation is a user's craft. It draws on the metamorphoses of type-form over time' and 'in craftwork, people can and do improve' [42]. Something more-ish in an artificial sense (and not culturally Moorish) is, therefore, a useful device for analysing the crafted luxury-impulses of within Marrakech's museum.

Drawing upon the theory of craft and its enduring human impulses relate to the attractive and transmissible ornament of brick-laying patterns. This includes the breathable brick screens (or brise-soleil) (Fig. 10). Enabling one to appreciate how organic materials relay alternative views relating to art and gastronomy. In other words, baking bricks links with cooking food as well as the desire to sense more of

Fig. 10 View of the YSL
Museum's brise-soleil from
the conference room. Photo:
A. Condello, 2019

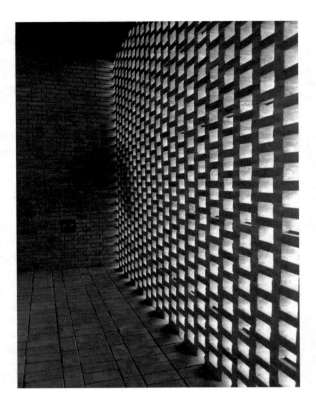

Fig. 10 View of the YSL Museum's brise-soleil from the conference room. Photo: A. Condello, 2019

something well-crafted, visually. According to Sennett, 'the long history of crafting clay shows three ways of becoming aroused consciously by materials, in altering, marking, or identifying them with ourselves. Each act has a rich inner structure: metamorphosis can occur through development of a type-form, combination of forms, or domain shift' [42, p. 144]. The biomorphic link to ancient Egyptian method of crafting the clay brick with a frog indentation thus improved brickmaking into a high-quality light product.

Crafting luxury, defined herein as 'the process of making a high-quality goods', which is a by-product of the desire to grapple more than what you intended pay, keeping in mind the respect for the environment, exposes the richness of reality. In the sense of the architectural object, the Yves Saint Laurent Museum thus provoked one to observe the origin and quality of the organic brickmaking process in crafting luxury, leading to unearthing biomimicry associations so as to explicate what 'more-ish' qualities imply.

4 'More-ish' Qualities

Originally, deriving from Berber (or *Amazigh*) culture, the term *'Mour'* was associated with more, something that arouses one's senses, which links with both luxury and architecture, and its quality of crafting. Wanton or benevolent desire of taste, for instance, would trickle into what French philosopher Immanuel Kant would put it as 'many kinds of agreeable sensations [which] are 'more-ish' or productively interested, [but] not *all* pleasures in the agreeable are' [43, p. 316]. Metaphorically speaking, 'more-ish,' is a colloquial term mostly associated with the desire for food. Perceived by some to be the world's first processed food, olives harvested in Northern Africa, for instance, in Libya, or the deep-fried *brik* pastries of Tunisia are considered to be 'more-ish'. Relating to tasty food, the term 'more-ish' in the Oxford English Dictionary makes reference to Jonathan Swift's 1738 Polite Conversation and also found earlier in the 1700s, but not in print [37, p. 3398]. Arguably, the word was detected in historical or contemporary accounts is neither associated with the Moors or *Mours* nor does it have any association with Moroccan cooking. Herein, one quality of more-ish implies the hunger to know more of sustainable Amazigh culture. Another quality is expressed as something sustainably beneficial within or outside architecture.

Architecturally, 'more-ish' pointed first towards actual tea-bricks. Then, it was associated with drinking a hot brew and meal services and certain spaces. In the case of 'brick teas' (which were first created in China about 1000 years ago then Siberia and subsequently in India), consisted of leaves, stalks and twigs were compressed [4, p. 367] flat into portable organic objects. Brick tea was also used 'as a form of currency' [20, p. 127]. In eighteenth-century England, Swift's account about a woman sipping tea is curious since the liquid became a popular drink and one desired more of it. This would have been the time when Twinning's Tea Room (ca 1706) was established in England where one would take tea plus consume cakes and savoury pastries in the afternoon. Meanwhile, earlier in the mid-seventeenth century tea was served publically through the coffeehouses as an indulgence. Eventually, tea became recognised as a builder's hot brew, so mouth-watering other people wanted some of the libation, if not, more of it too.

Tea was associated with dinner or 'meal service' as well, hence the whetting one's appetite, within a tea room. 'One example of a local tea experience is found in the distinctive culinary tradition that has evolved around the English afternoon tea. This is a meal service that usually consists of tea accompanied by sandwiches and sweets …. In some cultures, tea also plays another role in cuisines, either as a dish on its own or as a flavouring for local dishes' [20, p. 130]. One key modern architectural example of note is Charles Rennie Mackintosh's Willow Tearooms (1903) in Glasgow, Scotland, with its series of art tearooms. At that time, lavish materials were used to create Mackintosh's *Salon De Luxe*, but this case is more about the Arts and Crafts Movement rather than the reverse—crafts and art. Everything from the carpets to the cutlery was branded in Mackintosh's Salon De Luxe, perceived by some as an 'organic' composition. Meanwhile, in Morocco, the Maghrebi mint tea ceremony had already been established as a state-of-the-art custom, sitting in amongst modern Amazigh crafts, as well as throughout the rest of northern African countries. There, mint tea leaves were a ready-to-cook concoction to sip from the East in the open-air or within enclosed environments (Fig. 11). The simple custom of taking tea thus relates to the 'more-ish' qualities within designated areas that materialises the organic.

Amazigh historian Tawfiq al-Madani noted in a 1930s article that North Africa as a whole 'was isolated, in ancient times, from the rest of the world'. And that Imazighen 'lived in it a life of innocent simplicity, preserving some of their customs and traditions which they had brought from the East at an unknown date, or which they acquired from among the elements of Egyptian religion and culture' (quoted in [30]). On the one hand, East and West were at odds with one another, that is, 'an uncomfortable tension between the notion of the indigenous permanence and the constant pull of the East' [30] exists. On the other hand, Orientalist painter Eugene Delacroix categorised and described Moroccans 'as a thousand times near/closer to nature' [14, p. 18]. This magnetic pull between the East, North and West nature and Amazigh customs of Morocco questions what the secret constituent is, that is, the extra quality, especially when sensing you want to know more about something became 'more-ish' because of the excess of luxury crafts. Presumably, this extra

Fig. 11 'More-ish' mint tea scenario at the ES SAADI resort, Marrakech. Photo: A. Condello, 2019

quality was established well-before Delacroix—with Ibn Battuta's influence when he traded with China in the fourteenth century, especially in procuring handcrafted silks and 'tea-coins', rather than tea-bricks.

Other than brews and foodstuffs, something 'more-ish' is highly attractive, a quality that satisfies something for someone as well as a plead that urges someone to keep going back for more. Now, this chapter will turn to demonstrate the way that ideas relating to crafted luxury and sustainability, especially the brickwork patterns, are worked within Studio KO's Yves Saint Laurent Museum as a mechanism of more-ish qualities that explicate the trickery of the organic, by the camouflage of animal pelts made into permanent brick-pelts.

5 Brick-Trickery Craftwork at the Yves Saint Laurent Museum, an Organic Approach

Time-shifting within the contemporary architectural practice and the incorporation of traditional or forgotten craftsmanship has re-emerged through the making marble models was done in the Middle Ages by the master masons. At the two months conceptual stage of designing the Yves Saint Laurent Museum, which all-in-all took two years to construct in reality, Studio KO commissioned Atelier Misto founder Miza Mucciarelli from Italy to make a presentation model out of marble, reusing red porphyry (Fig. 12). Traditionally, imperial porphyry (found in Egypt's eastern desert and that was used to build sarcophagi for Roman emperors) is perceived to be a geologically unique rock, resembling the colour of purple blood. 'Blood, is what ties a soul to the earth, and it is also what produces and contains earthly memory' [18, p. 15]. Mucciarelli's time to craft the model thus presents the intellectual efforts of the architect as well as the imaginative reconstruction of excess luxury materials, resembling a 'haberdashery' of elegant off-cuts.

Located on the edge of the Majorelle Garden, the Yves Saint Laurent Museum, an austere but intricate brick-tomb with brick-pelts or facades for clothing, was inspired by the colours of the Marrakech desert-the local brick palate of the city and Amazigh culture. Created on behalf of the Majorelle Garden Foundation, the Yves Saint Laurent museum was managed by Studio KO's chief architect at the Marrakech office, Faycal Tiaiba.

Yves Saint Laurent Museum (Fig. 13) comprises an exhibition space, a temporary exhibition hall, bookshop, the Pierre Berge auditorium (a flexible space), library, conference room, café and restaurant and a fashion archive. The building plan is a strong sequence of circles, squares and rectilinear spaces. Horizontal layers of textured brickwork resemble a textile weave and the floor plan adapts and curiously represents Mohamed Chebba's painting entitled *Composition* (1968) (Fig. 14). The museum basement according to Director of the Yves Saint Laurent Museum Bjorn Dahlstrom 'comprises a state-of-the-art conservation department whereby hundreds of garments and accessories are stored with optimal conservation conditions in mind

Fig. 12 At Studio KO,
Marrakech. The architects
commissioned Atelier Misto
to craft the YSL museum
model using red porphyry.
Photo: A. Condello, 2019

to safeguard the textiles and preserve the items. The conservation rooms also comprise a quarantine section where other items, an atelier for dust removal and laboratory for restoration work' [36, pp. 146–147]. Strikingly, the museum's open-air drum courtyard presents a context of weaving in static action.

Apart from time being the essence of crafted luxury, visitors to the museum can experience not only the creation process, but also observe the reuse of ancient or forgotten crafts. For example, reusing marble off-cuts for the model and reconstituted marble and other stone chips into terrazzo products as in the museum's new pink facades are examples of sustainable crafted luxury. Based on the values of craftsmanship and recognised for their creative solutions for sustainable constructions, Moroccan building company Bymaro, a subsidiary of Bouygues Construction, managed the entire project. Bymaro's tradesmen also comprehensively respected the detailed technical requirements for the conservation of the Yves Saint Laurent collection.

A number of more-ish qualities are apparent within the museum's open-air drum courtyard, which demonstrates the revelry juncture between the crafting of terrazzo made with reused marble chips contrasted with the swathes of brick trickery. The exterior brick facades align strongly with aspects of Amazigh textiles as fixed looms. The

Fig. 13 Construction of the concrete open-air drum courtyard. Photo courtesy: F. Tiaiba, 2017 ©

Fig. 14 Mohamed Chebaa's
Composition, 1968. Photo:
A. Condello, 2019

courtyard is made entirely from local terracotta bricks, concrete and rust-coloured terrazzo with Moroccan stone chips. The bricks were sourced locally and the interiors are rendered with white plaster and painted black in one of the display areas. For ornamental purposes, zellige or glazed tile mosaics were laid within the atrium, paying tribute to the traditional green technique of firing Moroccan ceramics with a green coppery glaze. The museum's brick façades and brick screens are laid dramatically upon the concrete frame. Each elevation unleashes brick biomorphism—an unexpected disruptive patterning of the snakeskin—a spoil left behind and crafted out of fire-bricked clay—and cut palm fronds. As a collective, I call these camouflaged patterns 'brick-pelts' (Figs. 15 and 16). This kind of brick trickery created a series of permanent brick-pelts set upon static looms, which slightly shift when one manoeuvres around the museum by the bricks' protrusions. Showing the critical moment of visual deception when looking and moving away from any of the museum's façade, one appears to be on the verge of catching or being caught—attracting attention—most definitely a more-ish camouflage attribute.

Studio KO's chief architect at the Marrakech office, Faycal Tiaiba, explained to me how the two-storey Yves Saint Laurent Museum's brick and terrazzo courtyard drum was informed by American land-artist James Turrell's installation whose works focussed on spatial perception. One can definitely see traces of Turrell's Within

Fig. 15 View of the YSL museum's brick textile or brick patterned 'pelt'. Photo: A. Condello, 2019

Fig. 16 Rear brick-pelt of
the YSL museum. Photo
courtesy: F. Tiaiba, 2017 ©

Without and Sky space (2010) observatory at the National Gallery of Australia, in Canberra, within the courtyard drum. This is especially the case when gazing at the ground of the museum then being confronted by the swathes of staggered brickwork 'as if in a gigantic wind turbine' (Fig. 17), as mentioned in Benali's story at the beginning of this chapter, and then observing the strength of the sky. The archive is entombed within the interior and the entire collection houses more than a few thousand items of clothing, drawings and haute couture accessories. These technical premises ensure that the archive is climatically controlled and has the most advanced technology in Morocco. Other features include an atrium, a reflective rectangular pool and sustainable gardens planted with palms, papyrus (Fig. 18) and Barbary prickly pears (Fig. 19), all well-suited to the desert environment.

Studio KO's brick-trickery craftwork is all about preserving the cultural heritage presence of Marrakech's modernist buildings. For example, the Cinema Palace in rue Yougoslavie, Gueliz [31], built in 1926 (a replica of Cinema Eden in La Ciota in France) has a more-ish façade and open-air theatre. Studio KO has materialised within their design a rotation of the cinema's brick and rendered façade at 180 degrees and adapted, and camouflaged, it together with large samplers of Berber carpets built out of brick. Each of the five different types of brick textiles overhang the concrete structure as if they were permanent brick-pelts hanging vertically over

Fig. 17 Staggered
brickwork patterns within
the YSL museum's open-air
drum courtyard. Photo: A.
Condello, 2019

a riad's handrails. Olivier Marty notes, 'we prefer things that show they've been crafted by someone's hands' [11, p. 9]. Both of them favour the simplicity of things and 'learned the power of the hand '[11, p. 9]. In crafting luxury, Studio KO's Yves Saint Laurent Museum could be interpreted as preserving Marrakech's peculiar high-qualities of the Amazigh riad.

One month after Berge's unfortunate death, the Yves Saint Laurent Museum opened in October 2017. The inaugural exhibition was Jacques Majorelle's Morocco (2017–2018) [15]. The second one featured the work of Moroccan couturier Noureddine Amir [2] (Figs. 20 and 21). Influenced by global and local development, Noureddine Amir uses the medium of fashion to criticise ideas about Moroccaness. Amir 'turns to Morocco's cultural diversity represented in popular culture, street styles, rural cultural heritage, as well as the country's heterogeneous cultural past, including its African origins' [19, p. 142], which was originally crafted only by Amazigh women. 'Handicrafts can be viewed, then, as a major experimental instrument, as an industrial area open to new patterns and new designs that mass-production would be unable to create to the rigidity of its technical and manufacturing structure' [9, p. 580] crafted luxury, sustainably.

Amir's crafted garments use raw indigenous materials, such as raffia and wool, which are reworked in an innovative way to improve luxury craftsmanship. Amir's

Fig. 18 View of the YSL museum's reflective rectangular pool and restaurant. Photo: A. Condello, 2019

sculptural textile creations are both fashion and architecture from another time—they are handcrafted apparel and abode. El Aroussi writes, 'the textures, colours and shapes certainly remind us of Amazigh (aka Berber) constructions found in towns in southern regions of North Africa, from Morocco to Egypt' [12, p. 7]. Amir's work 'profoundly influences the austere lines of Berber costumes, he does not hesitate to introduce techniques ….. The interaction between Berber weavers and textile and fashion designers could well lead to new creations, geared into the third millennium' [28, p. 280]. Similar to Studio KO's textured textile method, Amir's crafted luxury creations 'awaken our instinctive drive to reconnect with nature and the earth' [12, p. 7]. Above all, these haute couture designers 'manufacture' in the proper sense of the term.

As haute couture architects, Studio KO express that:

Artisanat, a nice word that has the word art, although in English it contains craft. Morocco showed us the way of craftsmanship, the power of the hand, the vibration of a texture that can only happen when a man or a woman fabricates. It binds us with a past as well as the future. It creates the timeless. Probably working with artisans implies that you go towards the unknown, by incorporating elements that can never be completed specified, but all have to be tested, tried, and applied. There is the fear, the excitement of searching, of walking old paths that haven't been used for a while (quoted in [35, p. 30]).

Fig. 19 Sustainable gardens at the YSL Museum planted with Barbary cacti. Photo: A. Condello, 2019

6   Recrafting Amazigh Luxury

As a materialised product, high-quality function and experience, luxury became fastened to sustainable values. In terms of architecture, sustainability is about the endurance and longevity of a building material in a given context, but there are other more pressing concerns today as far as considering new creative solutions for luxurious constructions. Human impulses attracted the effort for the craftsperson to create intricate structures destined for the fashion mogul to house its luxurious materials and accoutrements in the form of a brick-tomb for tuning-into.

The crafting of architecture goes beyond what was created before. By respecting its past materiality, markings and traditions and the creation of new spaces with presence, craftsmanship inspires and attracts new forms of sustainable luxury—crafting luxury—that embraces both local and global transcultural traditions. For contemporary architecture to embrace and maintain the organic approach, the 'more-ish' qualities can certainly forgo the preservation of cultural convention. And instead accommodate the magical or rather more-ish properties of cultural consumption in desert built environments. Crafted luxury is therefore a more-ish functional excess to sustain lost cultural legacy.

Fig. 20 Moroccan couturier Noureddine Amir on the far left at his atelier. Photo courtesy: L. Broscatean, 2019

In terms of what is 'more-ish', it is a highly attractive quality that satiates something—a plead that urges someone to keep looking and searching for more layers. It does not metamorphose as a compulsion to know the secret element that draws someone to appreciate the improvisation of the crafted object and the experience of crafted luxury that take place in an enclosed or open-air spaces. More-ish qualities are linked with what is organic and this approach relates to the essence of something that bestows the experience of luxurious facades and spaces to be 'bewitched' by the sustainable elements within the landscape. Moreover, when observing Amazigh culture, the Yves Saint Laurent Museum has offered ample crafted objects for tourists and other visitors to immerse themselves to observe the 'more-ish' presence, now appreciated publically.

Drawing upon a system of more-ish interpretive qualities, the Yves Saint Laurent Museum showcases a fortified building with the earth and offers other bonding tactics for other contemporary artisans. The building is most certainly an immaculate and sophisticated case of the *finest*-crafted luxury. It also provides aspirational Amazigh cultural links to how craft and luxury merge as one within contemporary architecture. The structure also deliberately provides a crafted luxurious framework through which

Fig. 21 Author 'tuned-in' to touching one of Noureddine Amir's Amazigh luxury creations at his residence, Marrakech. Photo courtesy: L. Broscatean, 2019

ideas of sustainability and architecture can be massed and strengthened. Within architecture, more-ish qualities imply that you will receive more in context for what you pay for when exposed to the inevitable experience of more.

Since its opening, the Yves Saint Laurent Museum has gained worldwide recognition through the influx of photographs on Instagram and other social media. As a carefully crafted haute couture luxury-structure, the museum comprises a multitude of 'more-ish' qualities. Through Studio KO's integration of the ancient Amazigh products tradition, the qualities have been identified herein as a sustainable form of crafted luxury, and illustrate an organic architectural approach for the luxury sector. The Yves Saint Laurent Museum questions how the mind in-hand crafts Marrakech's original and modern Amazigh traces and the surrounding existing constructions within and outside the twisting layout of the medina.

As a whole, the city of Marrakech has thus regained its handicraft traditions by recrafting Amazigh luxury, which proves to be the organic approach to learn new technologies. For Branzi: 'artisan production recycles even the modern style, interchangeable with other formal models drawn from tradition; in any case, any ingredient of research or design is extraneous, and construction becomes synonymous with imitation' [8, p. 580]. In due course, brickmaking as a form of crafting luxury has cultivated a bonding or 'tuning-in' tactic, important in understanding the sustainable practice of Studio KO's architecture and their camouflaging and unveiling of 'more-ish'-Moroccan culture. By recrafting Amazigh luxury in future projects, indicates an enlivening organic approach, to be further transmitted by multiple artisanal impulses.

Studio KO architects operate as eloquent *artisanat*s and their Yves Saint Laurent Museum imparts to the public an invincible and fortitudinous silence—a tug towards recrafting Amazigh luxury.

Acknowledgements In Morocco, I thank Faycal Tiaiba from Studio KO Marrakech Office for his conversation on the Yves Saint Laurent Museum project and The Arts of Fashion Foundation for the opportunity to work within the YSL Museum's conference room as well as meeting Noureddine Amir at his atelier and residence.

References

1. Abu-Shams L, Gonzales-Vazquez A (2014) Juxtaposing time: an anthropology of multiple temporalities in Morocco. Revue des mondes musulmans et de la Mediterranee 136:33–58
2. Amir N (2018) The sculptural dresses: Noureddine Amir by Hamid Fardjad, Temporary Exhibition Catalogue 23 February – 22 April 2018, Marrakech. Yves Saint Laurent Museum, Marrakech
3. Aoudjehane HC, Avice G et al (2012) Tissint martian meteorite: a fresh look at the interior, surface and atmosphere of Mars. Science 338(6108):785–788
4. Baruah P (2011) Tea drinking: origin, perceptions, habits with special reference to Assam, its tribes, and role of Tocklai. Sci Cult 77(9–10):365–372
5. Becker CJ (2006) Amazigh arts in Morocco: women shaping berber identity. University of Texas Press, Austin
6. Benali A (2018) Looking at Mars in Marrakech. In: Adnan Y (ed) Marrakech Noir. Akashic Books, Brooklyn, New York
7. Berney KA, Ring T (eds) (1996) International dictionary of historic places, vol 4. Fitzroy Dearborn Publishers, Middle East and Africa, London
8. Branzi (2010) "The new handicrafts", from Hot house in The craft reader, edited by Glenn Adamson. Berg Publishers, Oxford; New York, pp 557–581
9. Branzi A (1984) "The new handicrafts", from hot house. In: Adamson G (ed) The craft reader (2010). Berg, Oxford, New York, pp 577–581
10. Cruickshank D (2015) The first cities. In: Hall W (ed) Brick. Phaidon Press Limited, London
11. Delavan T (2017) Essay in Studio KO, by Karl Fornier and Olivier Marty. Rizzoli International, New York
12. El Aroussi M (2018) 'The sculptural dresses: Noureddine Amir', curated by Hamid Fardjad. Temporary Exhibition Catalogue, Musee Yves Saint Laurent, Marrakech, 23 Feb–22 Apr
13. El Faiz M, Ruf T (2010) An introduction to the Khettara in Morocco: two contrasting cases. In: Schneier-Madanes G, Courel M-F (eds) Water and sustainability in arid regions: bridging the gap between the physical and social sciences. Springer, Dordrecht, Heidelberg, London, New York
14. El Aroussi M (2015) Visual arts in the Kingdom of Morocco, the Arab League's Educational Cultural and Scientific Organization
15. Fardjad H (2018) The sculptural dresses: Noureddine Amir, curated by H. Fardjad, Temporary exhibition catalogue 23 Feb–22 Apr 2018, Marrakech. Yves Saint Laurent Museum, Marrakech
16. Gardetti MA, Torres AL (2013) Sustainability in fashion and textiles: values, design, production and consumption. Greenleaf Publishing
17. Hampate Ba A (1976) African art: Where the hand has ears.' In: Adamson G (ed) The craft reader (2010). Berg, Oxford, New York, pp 379–385
18. Hansen C (2009) "Colors/ porphyry: blood from a stone", in Cabinet magazine, Spring, Issue 33

19. Jansen AM (2016) Defining Moroccanness: the aesthetics and politics of contemporary Moroccan fashion design. J Northern Afr Stud 21(1):132–147
20. Joliffe L (2004) The lure of tea; history, traditions and attractions. In: Hall M et al (eds) Food tourism around the world: development, management and markets. Routledge, London, pp 121–136
21. Kassia SC (2018) The golden thread. John Murray Publishers, London
22. Khalid M (2019) Urban restructuring, power and capitalism in the tourist city: contested terrains of Marrakech. Routledge, London
23. Kidder-Smith GE (1955) North African scrapbook: Morocco. Architect Rev 118(705):177–185
24. Koolhaas R (2002) Junkspace. In: Chung J, Inaba J et al (eds) Harvard school of design guide to shopping. Taschen GmbH, Koln
25. Koolhaas R (2001) Junkspace. https://www.readingdesign.org/junkspace. Accessed 27 Aug 2019
26. Kouskou T, Allouhi A et al (2015) Renewable energy potential and national policy directions for sustainable development in Morocco. Renew Sustain Energy Rev (47):46–57
27. Lamzah A (2008) The impact of the French protectorate on cultural heritage management in Morocco: the case of Marrakech. Ph.D. dissertation, University of Illinois at Urbana-Champaign, Illinois, USA
28. Marie-Rose R, Sorber F (2007) Berber costumes of Morocco: traditional patterns. ACR Edition, Paris
29. Marty O, Fornier K (2017) Studio KO. Rizzoli International, New York
30. McDougall J (2003) Myth and counter-myth: the 'Berber' as national signifiers in Algerian historiographies. Rad History Rev Spring 86:66–88
31. Meffre G, Delgado B (2012) Un Urbanisme experimental: Les villes nouvelles Marocaines (1912–1965), la Foundation Jardin Majorelle. Senso Unico Editions, Marrakech
32. Miller AV (2018) Tents, palaces, and 'imperial souvenirs': mobilizing cultural authority in the French protectorate of Morocco. J Middle East Africa 9(1):51–75
33. Moine R (2017) Saint Laurent on screen: fashion icon, doomed artist, or celebrity? Fashion Theory 21(6):733–748
34. Nicholas Claire (2014) Of texts and textiles…: colonial ethnography and contemporary Moroccan heritage. J North African Stud 19(3):390–412
35. Nicolin P (2018) How can I not think about you in Marrakech? In: Domus, no. 1026. https://www.domusweb.it/en/issues/2018/1026.html. Accessed 05 May 2019
36. O'Kelly E (2017) Less is more: the Yves Saint Laurent Museum is set to be a lesson in restrained elegance, Wallpaper*. https://www.wallpaper.com/author/emma-okelly/6. Accessed 05 May 2019
37. Partridge E (2006) The new Partridge dictionary of slang and unconventional English, edited by Terry Victor. Routledge, London
38. Pulice M (2006) Machines for making bricks in America, 1800–1850, The Chronicle of the early American industries associations, vol 59, no 2, pp 53–58
39. Ringer R (2019) Turning up the heat: Sydney's 19th century brickyards. Bricks to build a town in Australian Society for the Study of Labour History, https://www.labourhistory.org.au/hummer/the-hummer-vol-8-no-1-2012/brickyardssydney/. Accessed 30 Oct 2019
40. Risatti H (2007) A theory of craft: function and aesthetic expression. The University of North California Press, Chapel Hill
41. Saint Laurent Y (1998) Yves Saint Laurent and fashion photography, te Neues Publishing Company
42. Sennett R (2008) The craftsman. Allen Lane, Great Britain
43. Vandenabeele (2015) The sublime in art: Kant, the mannerist, and the matterist sublime. J Aesthetic Educ 49(3):32–49
44. Wiesing L (2019) A philosophy of luxury. Routledge, London

Annette Condello (Ph.D. the University of Western Australia), is Senior Lecturer in Architecture at the School of Design and the Built Environment, Curtin University in Perth, Australia. She was the School's former Director Graduate Research and Director International for two years. Annette was Visiting Professor at DICATAM, the University of Brescia, Italy, and held a Mexican Government Fellowship at UNAM, Mexico City. Recently, Annette participated at the World Urban Forum (wuf10) in Abu Dhabi. Annette has published *The Architecture of Luxury* (Routledge, 2014), co-edited a book with S. Lehmann *Sustainable Lina: Lina Bo Bardi's Adaptive Reuse Projects* (Springer Publishers, 2016), and *Pier Luigi Nervi and Australia: Outback Modernism* (Black Swan Press, 2017).

Printed in the United States
by Baker & Taylor Publisher Services